D0853198

GASEOUS ELECTRONICS
SOME APPLICATIONS

This volume includes five papers from the
25th Gaseous Electronics Conference held in London,
Ontario, Canada 17 - 20 October 1972

GASEOUS ELECTRONICS

SOME APPLICATIONS

Edited by

J. Wm. McGOWAN
and
P. K. JOHN

The University of Western Ontario, London, Canada

1974

NORTH-HOLLAND PUBLISHING COMPANY – AMSTERDAM • OXFORD
AMERICAN ELSEVIER PUBLISHING COMPANY, INC. – NEW YORK

North-Holland ISBN: 0 7204 0308 1
American Elsevier ISBN: 0 444 10777 0

PUBLISHERS:
NORTH-HOLLAND PUBLISHING COMPANY – AMSTERDAM
NORTH-HOLLAND PUBLISHING COMPANY, LTD. – OXFORD

SOLE DISTRIBUTORS FOR THE U.S.A. AND CANADA:
AMERICAN ELSEVIER PUBLISHING COMPANY, INC.
52 VANDERBILT AVENUE, NEW YORK, N.Y. 10017

PRINTED IN THE NETHERLANDS

EDITORS' NOTE

The twenty-fifth anniversary Gaseous Electronics Conference was held at the University of Western Ontario, London, Ontario, Canada, largely because of the support given our proposal by Professors Allis and Loeb. It was their imagination and willingness for change that allowed the Canadian students and scientists to personally participate in the Conference. The Conference was jointly sponsored by the Divisions of Electronic & Atomic Physics of The American Physical Society and of Atomic & Molecular Physics of the Canadian Association of Physicists. Major support came from the National Research Council of Canada, the City of London and The University of Western Ontario.

The papers which are presented in this volume are the invited talks given at the Anniversary Conference and as such are not meant to be exhaustive reviews of the field. Instead they reflect the development of gaseous electronics and focus upon fields both pure and applied where gaseous electronics plays a major role.

The credit for the preparation of this volume must go primarily to Marion Wilson who was responsible for the final typing of all manuscripts and to our many associates who have participated in reading and checking these manuscripts. In particular we would like to recognize two of our graduate students who have contributed greatly to the proofreading, Andrew Ng and John Box.

J. Wm. McGowan
Editors
P.K. John

"Photo by John F. Cook"

WILLIAM PHELPS ALLIS
Faculty member of physics, M.I.T. since 1931; B.S. Physics, M.I.T. 1923; D.Sc. Physics, U. Nancy, France, 1925; Legion of Merit, France 1945; Ass't. Secretary Scientific Affairs, NATO, France, 1962-64; Fellow, APS and AAAS; Chairman, Gaseous Electronics Conference 1949-1962, Honorary Chairman, 1964-present; Author of several books; Primary interest in free electrons in gases and plasma physics.

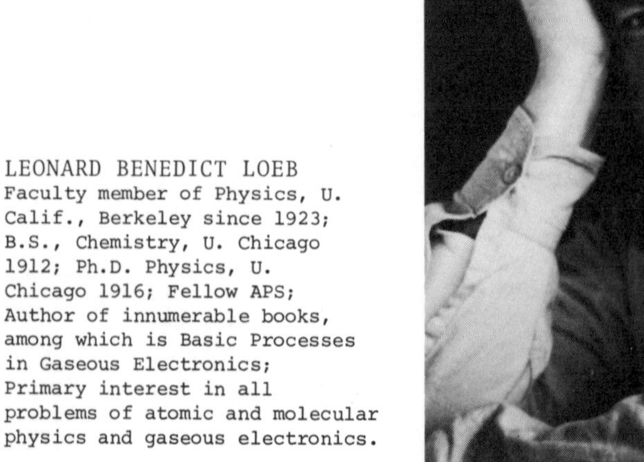

LEONARD BENEDICT LOEB
Faculty member of Physics, U. Calif., Berkeley since 1923; B.S., Chemistry, U. Chicago 1912; Ph.D. Physics, U. Chicago 1916; Fellow APS; Author of innumerable books, among which is Basic Processes in Gaseous Electronics; Primary interest in all problems of atomic and molecular physics and gaseous electronics.

DEDICATION

This volume, which commemorates the silver anniversary of the Gaseous Electronics Conferences, is dedicated to all who have been instrumental in the development of the field. In particular we identify two of its leaders, Professor William P. Allis, M.I.T., and Professor Leonard B. Loeb, Berkeley.

"Will" Allis was one of the prime movers among a small group of "founding fathers" of the Conference who has provided leadership and guidance as chairman or honorary chairman for most of its twenty-five year history. The continuing success of the conference is in a great measure the result of his interest and foresight at crucial points in its evolution.

Leonard Loeb has not only played a major role in developing the conference but has been instrumental in the continuing health of "Gaseous Electronics". In particular he has played an important role as rapporteur for the field through his many volumes and writings.

Those who have been privileged to work with either Allis or Loeb as students or colleagues can attest to their constant and unselfish helpfulness, their friendliness and continuing good humour, and their keen insight into many aspects of atomic and plasma physics.

It seems fitting that this tribute to Gaseous Electronics and in particular to two great scholars and human beings should take the form of a history of the conference and the field from which it sprang, together with the papers dealing with the impact of gaseous electronics on a number of basic and applied areas of physics.

M.A. Biondi

Physics Department
University of Pittsburgh

G.L. Weissler

Department of Physics
University of Southern California

CONTENTS

Gaseous Electronics, eds. J.Wm. McGowan and P.K. John

1

TWO HUNDRED YEARS B.C. (Before the Conference)

SANBORN C. BROWN

Massachusetts Institute of Technology
Cambridge, Massachusetts

To set the stage for a 25th anniversary celebration of the Gaseous Electronics Conference, I would like to comment briefly on the two hundred years before the Conference. To do this I will divide my talk into the Four Ages of Vacuum.

No Vacuum: where man-made vacua do not enter the picture.

Poor Vacuum: where the phenomena studied directly depend on the gas present.

Better Vacuum: where the phenomena depend only in a secondary way on the residual gas.

Good Vacuum: or the Era of the G.E.C. where the phenomena are independent of the residual gas.

First - the Age of the No Vacuum. It is only here that progress in our field has not been directly and primarily dependent on experimental techniques, where whole areas of research did not open up until particular advances had been made in vacuum technology.

It includes the host of natural phenomena which have been studied and reported on since history was recorded: the aurora, the glow of the night sky, St. Elmo's fire, sparks from cats and rugs, the conductivity of flames, and that incredibly lethal experiment which Benjamin Franklin survived that rainy afternoon in 1752 but which has led to the suicide of more than 300 unlucky experimenters since then, whose impedance to ground was not so accidentally fortunate as than of Uncle Ben's. But even in the No Vacuum Age, progress was intimately tied up with apparatus discoveries. Great steps forward were made after the discoveryof the Leyden jar -- which incidentally was not discovered in Leyden but in Pomerania by Ewald Georg von Kleist in 1745. It was called the Leyden jar because Professor Pieter van Musschenbroek of Leyden almost killed himself experimenting with it and ended his published account with the statement that he would not receive such a shock again even if offered the whole kingdom of France. And for such "unworthy sentiments" he was severely rebuked by Joseph Priestly who in his History of Electricity compared the "cowardly professor" with the "truly philosophic heroism" of one Georg Wilhelm Richmann, a Swedish physicist who was one of the first unfortunate victims of an attempt to reproduce Franklin's kite experiment.

Sir Humphry Davy's discovery of the electric arc in 1821 had to wait for a sufficiently big battery to provide the necessary starting voltage, and high-voltage steady-state experiments were not really possible until Ruhmkorff's induction coil came into general use in the 1850's. Many famous names besides Franklin and Davy are associated with early atmospheric pressure experiments.

An interesting account is given by Volta (Phil. Trans. 72 275 (1782)) of some experiments he made in company with Lavoisier and Laplace. They found that electricity was obtained from the evaporation of water, from the combustion of coal, and from the effervescence of iron-filings in dilute sulphuric acid. The method was to place some burning charcoal on an insulated place and to throw water on the charcoal. The alteration in the potential of the plate was attributed to the evaporation of the water.

Coulomb in 1785 studied the charge leakage of insulated charged bodies and accounted for it by convective discharge through air, and both Ernest Rutherford and J.J. Thomson measured the effects of conducting gases at atmospheric pressure ionized by radioactive disintegration.

Many of the concepts which today we use as a matter of course in gaseous electronics are, as one might expect, from other fields of physics and can properly be discussed as No Vacuum phenomena. It was in 1860 in talking about ordinary unionized gases that Maxwell introduced the concept of mean free path. The Maxwell-Boltzmann distribution was first used for electrolytic ions by Nernst in 1889 and I suppose most of you know that what we now regard as our own particular property, the Debye length, came from the work in the middle 1920's of Debye, Hükel, and Onsager on shielding lengths in liquid dielectrics.

Perhaps one of the nicest stories in recent gas discharge history is the story of the Lord Rayleighs, and I have heard Rayleigh's name many times during this Conference. It was in 1906, at a time when Lord Kelvin and Johnstone Stoney were scrapping about whether to call the light electrically negative particle an ELECTRION or an ELECTRON (Stoney won) when the attention of many physicists was turned to explaining spectral lines. J.J. Thomson had proposed the watermelon atom model, where electrons were imbedded in a positive jelly, as seeds in a watermelon, and Lord Rayleigh, calculated what we now call the "plasma frequency" to predict the Balmer spectrum for hydrogen. True, it could not predict any other atoms, and Rutherford's hard-core nucleus won out in the end, but the formula for the frequency stayed with us and by quite an odd coincidence Rayleigh's original paper ended up in the Air Force Cambridge Research Laboratories in Lexington, Massachusetts. It is quite a nice story which some of you have heard, I'm sure.

A few years ago the Air Force Laboratories decided that it was most important to study the details of the glow of the night sky and got authorization to spend several million dollars on an expedition above the arctic circle to gather data. While they were getting ready to go, somebody happened to wonder where all the data was upon which the Fourth Lord Rayleigh spent his life. The Rayleigh of the airglow was the son of the famous Lord Rayleigh, and from the son's own observation at his country estate of Terling in Essex he had contributed steadily for thirty years papers on air glow. John Howard, the Cambridge Air Force Laboratories' Chief Scientist, got so intrigued that he went calling on the present Fifth Lord Rayleigh and soon discovered that the data they needed had already been accumulated and after suitable negotiations arranged to buy all the data books at a tiny fraction of the cost of an expedition. In due course, box after box of paper arrived in Lexington, Massachusetts, from England, and when they opened them up they discovered that they had bought not only the Fourth Lord Rayleigh's papers, but all those of John William Strutt -- the famous Third Lord Rayleigh's papers as well!

Before I leave the Rayleighs I must point out how badly history has treated the son of the famous father. Robert Strutt, the Fourth Lord Rayleigh, is almost unknown. I'll bet not more than a couple of people in this room have ever heard of him, yet I would also bet that nobody in this room can match his productivity in the field of gaseous electronics. Between the turn of the century and the First World War he published over 300 papers on air glow and gaseous discharge theory.

But now I must turn from the era of No Vacuum to the era of the Poor Vacuum. This era was ushered in by an itinerant glassblower named Geissler who eventually became a professor of physics at the University of Bonn. In 1858 Geissler developed a technique of fusing platinum electrodes directly into glass, and he wandered from college to college throughout Europe earning his living by making "Geissler" tubes which were the showpieces of late nineteenth century demonstration physics lectures.

These tubes were called "vacuum" tubes although up to 1880 they were evacuated by piston pumps whose ultimate limit was about 1/4 of a Torr. In 1874 McLeod introduced his famous gauge and shortly thereafter Toepler devised a pump which worked like a McLeod gauge except the top measuring bore was attached to the tube to be evacuated so that by patient and repeated raisings and lowerings of the mercury column the pressure in the tube could be lowered. Those happy graduate students who nowadays turn on their pumps and go off to class or home to their wives and kids have no concept of what it was like to evacuate a tube at the turn of this century. My own father was a graduate student of Townsend in Oxford, getting his graduate degree in 1906 by studying the fall in potential between breakdown and steady state in a glow discharge, and many is the time that he has described to me the agonizing hour after hour of operating a Toepler pump even to start an experiment.

But the Age of the Poor Vacuum was one of tremendous progress dominated by Hittorf and Crookes.

Two people in such different circumstances as Hittorf and Crookes are hard to imagine. To quote from K.K. Darrow's The Renaissance of Physics

"Hittorf betook himself to a small town half starved university where he became professor of physics and chemistry both with a heavy teaching load and no appropriation for research."

"Crookes in England was a man as fortunate in wealth and leisure as Hittorf was strained and skimped....He seems to have begun after Hittorf, and to have seen rather deeper; but this is one of the painful questions of 'priority' which has become particularly embittered."

But between the two of them they led the field up to the time of J.J. Thomson, Rutherford, Zeleny, Langevin, and Röntgen whose accidental discovery of X-rays as he was playing with one of Geissler's tubes ushered in a whole new physics. The parallel lives of Crookes and Hittorf would make a story all to itself. Crookes a wealthy aristocratic Englishman seemed always successful with his carefully pointed waxed mustache ordering about his crew of technicians and glassblowers. Hittorf, the always half-starved overworked German, laboriously building his own equipment when he could sneak away from his overload of teaching, was always having trouble. Trouble like the time when he was trying to determine if there was an ultimate length to the positive column. Week after week his discharge tube grew as he added meter after meter, sealing it off, pumping it down, and testing it as he pushed his Ruhmkorff induction coil to higher and higher voltage. His tube went all the way across the room, turned and came back, turned again until his whole laboratory seemed full of thin glass tubing. It was summer and as he sweated away with his Toepler pump he opened the windows to make it bearable. Suddenly from outside came the howl of a pack of dogs in full pursuit and flying through the window came a terrified cat to land feet spread for the impact in the middle of the weeks and weeks of labor. "Until an unfortunate accident terminated my experiment," Hittorf wrote, "the positive column appears to extend without limit."

It was Crookes who clearly viewed the ionized state as a unique condition of matter. To an audience at a meeting of the British Association for the Advancement of Science in 1879 he said -- and let me quote, since it would never get by modern referees: "So distinct are these phenomena from anything which occurs in air or gas at the ordinary tension, that we are led to assume that we are here brought face to face with Matter in a Fourth state or condition, a condition so far removed from the state of gas as a gas is from a liquid. In studying this Fourth state of Matter we seem at length to have within our grasp and obedient to our control the little indivisible particles which with good warrant are supposed to constitute the physical basis of the universe....We have actually touched the border land where Matter and Force seem to merge into one another, the shadowy realm between Known and Unknown, which for me has always had peculiar temptations. I venture to think that the greatest scientific problems of the future will find their solution in this Border Land, and even beyond; here, it seems to me, lie Ultimate Realities, subtle, far-reaching, wonderful."

Hittorf and Crookes, of course, were not alone in the field. De la Rue and Müller in 1880 worked out the details of what we now call Paschen curves, J.J. Thomson and E. Rutherford worked on the conductivity of the steady-state discharges, Zeleny and Langevin studied ion mobility, Oliver Lodge in 1897 defined what we now call the cyclotron frequency from the work of Helmholz and Oersted, Townsend derived the transport equations in 1899, and Lorentz worked out the energy gain equations for charged particles in an electric field, at the same time that Einstein derived the equations for free diffusion.

The first text in gas discharge physics appeared as a chapter in a book by Clerk Maxwell's student J.J. Thomson who introduced it with: "The importance which Maxwell attached to the study of the phenomena attending the passage of electricity through gases, as well as the fact that there is no summary in English textbooks of the very extensive literature on this subject, lead me to think that a short account of recent researches on this kind of electric discharge may not be out of place in this volume."

The discoveries of Röntgen and Becquerel initiated the new physics of atomic structure and radioactivity and the passage of electricity through gases grew up to be a distinct field of its own when J.S. Townsend left Cambridge in 1900 and made Oxford the center of gas discharge research.

So also came the era of the Better Vacuum which is so well-covered in the textbooks we all have studied that I need not dwell on the details. Gaede invented his mechanical rotary oil pump in 1907, and in 1915 he invented the mercury diffusion pump, which became so much used that one could also call this era the age of the mercury containment. During the First World War our own Leonard Loeb got his degree and his tremendous energy and productivity did much to keep the field alive between the wars. He was not alone, of course, and I must mention one incident which changed the very language we use to characterize our field. I once asked Lewi Tonks to write down for me the story of Langmuir's introduction of the word "plasma" and what he wrote to me was the following:

"The circumstances, as recalled after four decades, were something like this: Langmuir came into the laboratory room and said, 'Say Tonks, I'm looking for a name. In these gas discharges we call the region in the immediate neighborhood of the wall or an electrode a 'sheath' and that seems to be quite appropriate. But what should we call the main part of the discharge? The conductivity is high and you can't apply a potential difference to it like you can to a sheath - it all is taken up by the sheaths. And there is complete space-charge neutralization. I don't want to invent a word, but it must be descriptive of this kind of a region as distinct from a sheath. What do you suggest?'

"My reply had an element of the classical: 'I'll think about it, Dr. Langmuir.'

"The next day L. breezed in and announced, 'I know what we'll call it' we'll call it the 'plasma'. The image of blood plasma immediately came to mind; I think L. even mentioned blood."

But Langmuir was not one to be deterred by other people's concepts even in the definition of words and as a good student of Greek he had studied his etymology. He was impressed by the obvious characteristic that the glowing discharge molded itself to any shape into which the tube was formed, so he chose the Greek word πλάσμα which means "to mold" and the word came into general use when research problems of controlled thermonuclear fusion involved large numbers of physicists who began to refer to their working "fluid" as a Langmuir plasma.

As I have mentioned, most of you know the history of the Better Vacuum Era and the Good Vacuum Era and the Good Vacuum Era is what Will Allis is going to talk about (chapter 2), so I want to end with just one more comment again about language. The Good Vacuum Era can also be designated the Gaseous Electronics Era and the Better Vacuum Era as the Gas Discharge Era. But those of us who started our professional lives in the Gas Discharge Era were remarkably sensitive to other possibilities of interpreting what we were studying. My wife almost didn't say "yes" to me in 1939 when she contemplated having to tell her friends what my research was called, and it was only the marvels of electronics which came out of the Second World War which saved us from tonight celebrating the 25th anniversary of the Gxxx Dxxx Conference instead of the G.E.C.

I am sure that part of my own sensitivity comes from my studies of the 18th century physics well-illustrated by a famous print by John Gillray of a gas discharge experiment at the Royal Institution in London conducted by Thomas Young of optical diffraction fame, Humphry Davy, the arc man, and my own particular friend, Count Rumford, illustrated in the following, my only slide.

From a caricature by James Gillray (1802)

Gaseous Electronics, eds. J. Wm. McGowan and P.K. John

TWENTY-FIVE YEARS OF GASEOUS ELECTRONICS

WILLIAM P. ALLIS

Department of Physics
Massachusetts Institute of Technology
Cambridge, Massachusetts

Next June I will attend my 50th class reunion at the Massachusetts Institute of Technology and this is the 25th Gaseous Electronics Conference. So the G.E.C. covers half my active career, and it is the better half, for it is at these conferences that I have obtained ideas in discussions with others and in which I have been able to test my ideas in discussions with you. It is this interaction which is, to my mind, the main purpose of the G.E.C. rather than the inherent right to present a paper.

These conferences grew out of the Nottingham Conferences, which were excellent conferences because of the discussions they stimulated, and unlimited discussion was allowed. Unfortunately, there were a limited number of days: three days, and therefore, discussion being unlimited, some people got squeezed out at the end of the conferences. This was the case for Gaseous Electronics. We got squeezed out on the Saturday sessions like toothpaste out of a tube. Several of us got together, Dan Alpert, Sandy Brown, Leon Fisher, John Hornbeck, Julius Molnar, some others, and I to seek a similar but better regulated conference in which gas discharges would have time to be discussed, and we organized the Gas Discharge Conference in 1947 at Brookhaven. This conference was held a Thursday, Friday and Saturday; there were 230 in attendance; 38 papers were presented, and 25 minutes per paper were available, including presentation and discussion. At this conference Margenau spoke on velocity distributions, Dieke and Donohue on striations; Holstein on imprisonment of radiation. This paper by Holstein is frequently quoted; it has been quoted several times at this conference, but very few people read it. It needs to be redone because line profiles and distributions have changed since it was first written, but nobody dares undertake this task. Ted (Holstein) was a prominent member of our earlier conferences, characterized by comments or questions on practically every paper that was presented. We lost him to the solid state but welcome him back to this meeting tonight.

The second meeting was sponsored by Westinghouse and held at the Mellon Institute in Pittsburgh. Better vacua were coming in, and positive ions were gradually coming out of the muck. Positive ions had always been in clusters, and at last we were getting clean ions, and there was much confusion over the fact that singly ionized rare gas atoms generally had a lower mobility than the corresponding molecular ions. It was at first not understood why the lighter particles should go more slowly.

At the Mellon conference, Emeleus spoke on "plasma oscillations", which became a by-word to explain any spurious result. And the first papers of a long series on the retrograde motion of the cathode spot were given. After this conference we sought respectability by seeking sponsorship by the American Physical Society, but we wished it on our own terms: we wanted to keep the right of reject ing papers arbitrarily and giving different time allotments to different speakers according to what we thought would be the need. In other words, we wished to keep the best conference for the audience to discuss physics, not just for the speaker to have a right to present his ideas.

The next meeting, sponsored by Bell Labs and held at the Barbizon Plaza in New York, was sponsored by the Division of Electron Physics as well. In spite of our youth, the Physical Society had accepted our terms. At the Bell Lab meeting Sanborn Brown spoke on microwave breakdown, and the famous paper on the mercury argon discharge by Kenty, Easley, and Barnes, the first detailed calculation of excitation and de-excitation rates and stepwise ionization was presented at this session.

The next conference at the Schenectady Laboratory of General Electric Company was famous largely for its banquet, at which all the celebrated names in gas discharges were present, Hull, Langmuir, Dushman, and Tonks. Dushman was particularly happy at this conference because he had just finished writing 300 pages about nothing, or as close to nothing as he could get.

The fifth conference was sponsored by RCA at Princeton. Rose and Allis gave a paper on the transition to ambipolar diffusion, a subject also discussed at this conference. And Kruskal presented a paper on the instabilities of plasmas.

The sixth, sponsored by the Office of Naval Research was in Washington, D.C. At this conference, Looney and Brown presented their paper on plasma oscillations, and Bennett his famous paper on self-focusing beams.

The seventh conference was sponsored by New York University, and had the famous Symposium on Breakdown in which Fisher and Loeb were pitted against Llewellyn-Jones: Streamer breakdown vs. Townsend Avalanche breakdown. Loeb is unfortunately unwell, and absent, for the first time, from one of these conferences. His continued determination to work on sparks and streamers is continued now by Goldmann who is here from Paris, but unfortunately not speaking: there was no more time available. Progress in this field has been largely set by the gradual improvement of timing circuits. After the meeting at New York University we were greeted by the results of Hurricane Hazel, which had blown down many trees across the roads, as we returned to our homes.

At the eighth conference, again at General Electric, we heard the paper by Pack and Phelps on drift velocities of electrons in molecular gases. This was the first attempt to derive from mobility measurements the cross sections for excitation of rotation and vibration in molecules. Due to the poor resolution of low energy electron beams there was, at the time, no other way. It is a classic paper, although now outdated.

The next conference at Westinghouse lasted three and a half days because, to our regular fare, we added a Sherwood Symposium. The Sherwood program (on Controlled Thermonuclear Reactions) had just been declassified, and Freeman, Kruskal, Post, Simon and Tuck spoke about the recently declassified work. Also at this conference Buchsbaum talked about microwave measurements of electron densities above plasma resonance, and Margenau on Balmer lines as indicators of electron densities and temperatures. Both these were early papers on diagnostics which could be made without introducing a solid body into the plasma. It was the beginning of great improvements in diagnostic methods.

At the next conference at M.I.T., I missed a great opportunity. I spoke at the banquet, but had not heard that Sputnik would be circling overhead at the very time I spoke. What an occasion to announce: Gentlemen, you are now within the orbit of a Russian rocket!

Next comes the second conference at Bell Labs. This time a four-day conference, with 62 papers. At this conference Schulz spoke about the trapped electron method, a method which improved the accuracy of electron energy measurements by an order of magnitude. It has improved again by another order of magnitude since, and this is another great change in gaseous electronics.

The twelfth conference was at the National Bureau of Standards, and here 85 papers were submitted. This created a crisis: we could not present 85 papers and maintain discussion, so we decided that we had to reject a large number, and we chose all papers having to do with surfaces as being inappropriate; gaseous electronics being really restricted to the gas. As a result, 38 papers were rejected, leaving the conference with 47 papers. Of these twenty were on collisions, thanks to the efforts of Branscomb. At this conference, Hirshfield and Bekefi talked about cyclotron harmonic radiation, again a very good diagnostic tool not requiring the introduction of a solid body into the plasma. We could now learn much about what was going on in a plasma by simply observing radiations that came out.

After this conference, arcs left us because they were attached to their cathode spot. Sherwood left us, because their field was too big and specialized, and formed the Division of Plasma Physics; and collisions also left us, to form the Collision Conference. Also, we went west for the first time to the U.S. Naval Post-Graduate School at Monterey. Although the number of papers was diminished by this move we still had as many as we wanted, and they were good. This conference lasted three and a half days. At this conference thermionic converters were first discussed.

The next conference, the third at General Electric, Ali Javan discussed lasers, but not by that name, and it was a few conferences later before anybody else introduced laser discharges into our conferences.

After this conference, I resigned as chairman because I would be out of the country and I missed the next two conferences. As I had been chairman for fourteen conferences, I think many people had been wondering how to get rid of me politely, and this was a good opportunity.

When I came back for the seventeenth conference, sponsored by the Signal Corps at Fort Monmouth, the number of papers had greatly increased and the time-per-paper decreased, even though simultaneous sessions had been introduced. At the Fort Monmouth conference Schulz spoke about resonances in elastic cross sections. Theory had failed to predict these resonances and indeed theory had also failed by orders of magnitude to calculate cross sections for rotational and vibrational excitation. Even though all the basic equations are known, there is still need for much theoretical work in atomic physics. Schulz's measurements again involved a large improvement in the measurement of very small currents, as he was using crossed beams of atoms and electrons.

From here on, I am unable to select outstanding papers as I have done up to this conference. Presentation had become too compressed; abstracts were too short; and anyway, due to simultaneous session, I usually failed to select the proper session to hear the outstanding paper. Eighty-five papers seem to represent an intellectual absorption limit.

The next conference was at Minnesota, and after that we went South to Georgia Institute of Technology. We had been invited by Duke several years earlier and everything had seemed very attractive until we looked into the question of hotels. At that time all hotels in Raleigh were segregated, so we regretfully had had to refuse the kind invitation by Duke. Now, with desegregation in the South, we could accept a meeting at Georgia Tech. At this meeting, Carl Kenty gave his last paper. He had attended all our conferences, had spoken at many of them and had also presented several striking experimental demonstrations. Many of us remember his red and white discharges due to cataphoresis and his strongly constricted filamentary discharge in a nitrogen and helium mixture.

Then we went to Lockheed, at which there were 111 papers, and the next one at JILA in Boulder, at which there were 116. Then to Oak Ridge where I have the attendance record: there were 360 attending, not a great increase from the original attendance at Brookhaven; however, there were 160 papers, a four-fold increase. At Brookhaven there were roughly six people attending per paper presented; at Oak Ridge, two people per paper presented. This drastic increase in the number of papers per person attending is due to the "no paper, no travel money" policy of most laboratories, and is one of the greatest obstacles which we have to organizing a conference in which discussion can be free and easy, and yet keep the whole conference down to a reasonable time. Of the people attending at Oak Ridge, five percent were at the first two meetings at Brookhaven or at the Mellon Institute. There were at this conference six arc sessions, two laser sessions, six panels, and seven regular gaseous electronics sessions.

Then we went to Hartford at United Aircraft, at which there were seven arc sessions and seventeen gaseous electronics sessions. This conference was famous for its interruption -- at one point Bob Bullis made a frantic announcement, saying, "You can't imagine what has happened to us. Agnew is going to speak here tonight, and he has pre-empted one of our conference rooms."

Then we went to the University of Florida, at Gainesville. There were eight arc sessions, five laser ones, sixteen gaseous electronics, and only 14.8 minutes available per paper. This is the low point in our time available, and we became obliged to eliminate the padding by papers which the author himself would frequently prefer not to give.

Now, at the University of Western Ontario, (London, Canada), the number of papers has increased even more, but thanks to very good management, and the use of predawn darkness and evening sessions, we are at 16.2 minutes per paper, with triple sessions.

I have told you what the conferences have been; perhaps you would like to know who are the G.E.C. Fred Biondi has presented papers at eighteen conferences; Phelps at seventeen; Leonard Loeb has spoken at thirteen conferences; Chanin at twelve; Kenty at ten; Fischer, Goldstein, Schulz at nine; These have been our leaders and if this order is not quite correct it is because there are two conferences of which I do not have the programs, and also because I may well have missed some names in looking through the lists of authors. If so, I must apologize to them.

A subcommittee has now been formed to consider the overcrowding of our conferences. They will recommend something, and it will inevitably hurt some of you. Should you suffer at the next conference I hope this will be compensated at some later meeting and if you will bear with us for some 25 conferences I am sure statistics will even out and you will come out way ahead. Or maybe you will name a different subcommittee.

Gaseous Electronics, eds. J.Wm. McGowan and P.K. John

3

APPLICATIONS OF GASEOUS ELECTRONICS TO LASER TECHNOLOGY †

A. V. PHELPS *

Joint Institute for Laboratory Astrophysics
National Bureau of Standards and
University of Colorado, Boulder, Colorado

I. INTRODUCTION

The current surge of interest in gas discharge lasers and in flash-lamp pumped solid state lasers constitutes an important new area for application of the research results which have been the subject of Gaseous Electronics Conferences for the past 25 years. In this brief review examples of these applications of research to only four of the currently important types of gas discharge lasers will be cited. Just as it is necessary to omit discussion of many of the important types of gas lasers, it will be possible to include only a fraction of the technologically important aspects of atomic and molecular physics, transport and distribution function studies and discharge phenomenology which make up this Conference. Applications of the results of Gaseous Electronics research to the high pressure flash-lamps used for pumping solid state lasers is more properly the subject of the following chapter. It should be noted that several very thorough reviews of gas discharge lasers are available (1-4).

II. He-Ne LASER

The helium-neon laser (5) developed by Javan, Bennett, and Herriott was the first successful laser utilizing a gas discharge as a working medium. The term level diagram appropriate to this laser is shown in Fig. 1. The condition for gas laser operation is that one produce sufficient stimulated emission during the passage of light through a gas containing excited atoms or molecules to overcome losses in the optical system. An excess of stimulated emission over absorption by the atoms requires that there be a population inversion; that is, the upper state density divided by its statistical weight must be higher than the lower state density divided by its statistical weight. In the He-Ne laser we are talking about radiative transitions from the 3s and 2s (Paschen notation) states of neon down to the 3p and 2p states,(1) and we need to consider the various excitation and de-excitation processes that result in population and depopulation of these levels. The principal excitation process for the neon is excitation transfer from helium atoms in the 2^3S and 2^1S metastable levels. Therefore one is concerned with the mechanisms for producing the helium metastables, the process of transfer to the neon and any processes that affect the populations of these levels. The properties of the helium metastables (6) have been discussed extensively at these meetings including the present one. The production

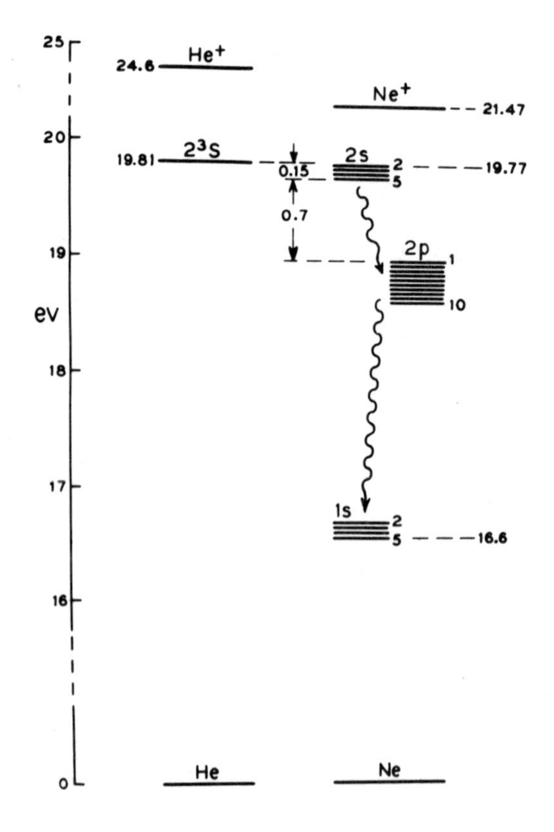

FIG. 1. Term level diagrams for the helium and neon atoms showing the levels of importance in the HeNe laser. (5)

and destruction of excited helium by electron impact and heavy particle impact and the collisions to and from nearby levels and the diffusion of the helium metastable to absorbing boundaries have also received considerable attention. The excitation transfer process in the helium-neon system has been studied primarily by those who are interested in laser operation (1,2). One of the problems is that the energy filters down to the metastable states and stays around for a long enough time so that there can be electron impact excitation back up to the lower laser level. Thus, we get into the question of what are the properties of neon metastables(6) which we've also discussed in great detail over the years.

III. He-Zn LASER

The He-Ne laser is sort of the archetype of an energy-exchange laser, if you want to use that terminology. There are other types which involve charged particles. The work of Silfvast(7) and others has shown that laser operation can be obtained for a number of elements by adding metal atoms to helium. Figure 2 shows a term level diagram taken from Webb, Turner, Smith, and Green(8), who carried out experiments designed to elucidate some of the population processes in a helium-zinc laser. They set up a flowing afterglow experiment with a microwave discharge to produce excited and ionized helium. They then added zinc downstream and looked at the radiation that was produced. The data are shown for zinc because they illustrate several of the processes that can occur. If a helium ion collides with a zinc atom in the ground state, there is sufficient energy not only to ionize the zinc but also to excite it to a fairly high level. This process is then charge transfer with simultaneous excitation. The open arrows in Fig. 2 are intended to indicate the transitions which were observed and attributed to this process. The width of the arrow is a measure of the intensity of the radiation. Those transitions which had at that time been observed to undergo laser action are indicated in boxes. The helium metastables do not have enough energy to produce these higher states, but they can ionize the zinc in a Penning process and produce the lower excited states of the ion. The radiation resulting from ionization and excitation by metastables is indicated by the cross-hatched arrows. The metal ion lasers utilize a second type of excitation transfer laser involving atomic species, and of course, the slightly different

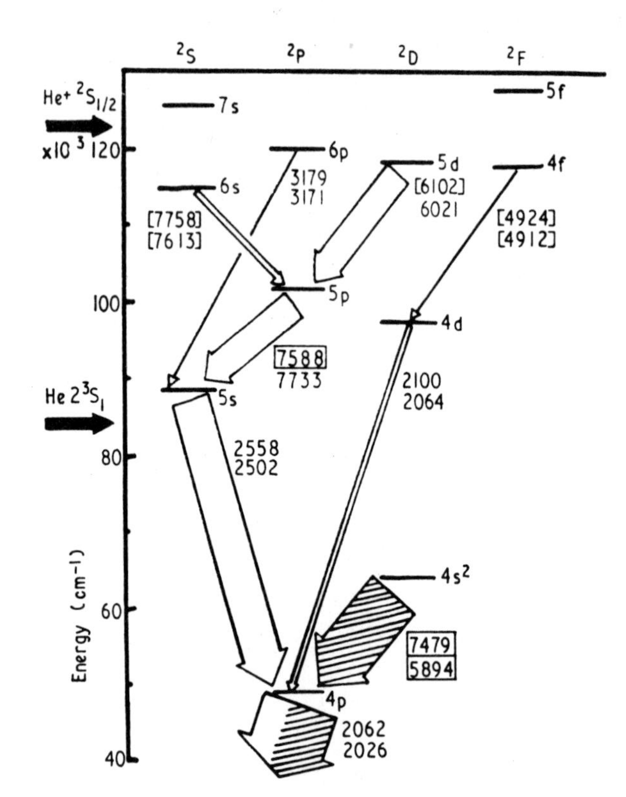

FIG. 2. Term level diagram for singly ionized zinc showing energies of the He 2^3S metastable, the He^+ ion and levels observed when Zn atoms are added to a He afterglow. (8).

processes of charge transfer and Penning ionization. Again, these are subjects that have been discussed in much detail at these conferences.

IV. CO_2 LASER

A third type of excitation transfer laser involves molecular systems which will be discussed in a little more detail now. Figure 3 shows the term level diagram appropriate to the N_2-CO_2 system. In this system the laser transition that we're interested in occurs between

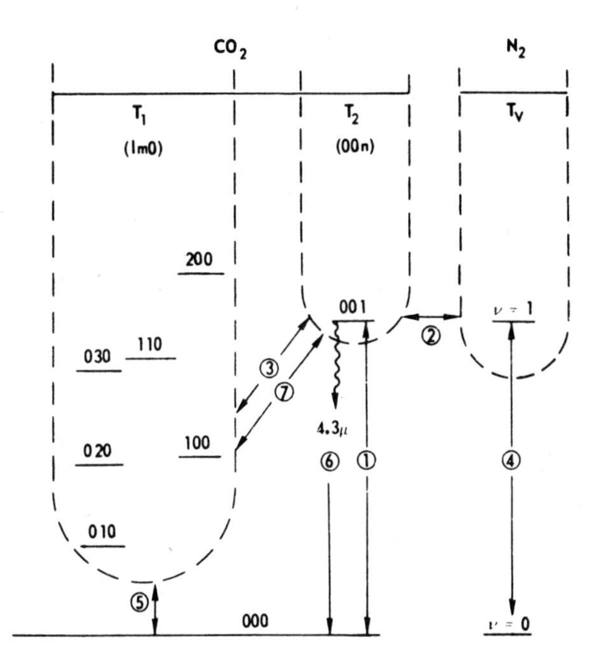

FIG. 3. Term level diagrams for the lower vibrational levels of CO_2 and N_2 showing grouping of levels for modeling and important radiative and collision induced transitions(4).

the 001 level of CO_2 and the 100 level of CO_2. The excitation transfer occurs between the first vibrationally excited state of N_2, shown on the right, and the 001 level of CO_2. It is convenient in analyses of this system to divide the vibrational modes of CO_2 in two groups. The upper laser level belongs to the asymmetric stretch mode. The relative populations of the vibrational states of this mode can usually be described by a temperature. One can combine the levels of the bending mode and the symmetric stretch mode into what is called the mixed mode. Because of the close coupling between the 020 and 100 levels one can usually assign a single vibrational temperature to all of the vibrational levels of this mixed mode. Since we are concerned with developing models which will allow us to predict the population inversion between 001 and 100 vibrational levels we are concerned with excitation and deexcitation processes by electrons(9,10), with collisional processes(11) connecting the vibrational systems of CO_2 and N_2, and with collisional relaxation to the ground state(11).

The only radiative process of importance is that of stimulated emission.

A. Vibrational Excitation

Since the excitation of the upper laser level of CO_2 by electrons is not a particularly favored process in pure CO_2, it has been found desirable to add N_2 to the system and take advantage of the high efficiency of electron excitation in N_2 and the close resonance between the first vibrational state of nitrogen and the upper laser level(3). Figure 4 shows the original measurements of the cross section for

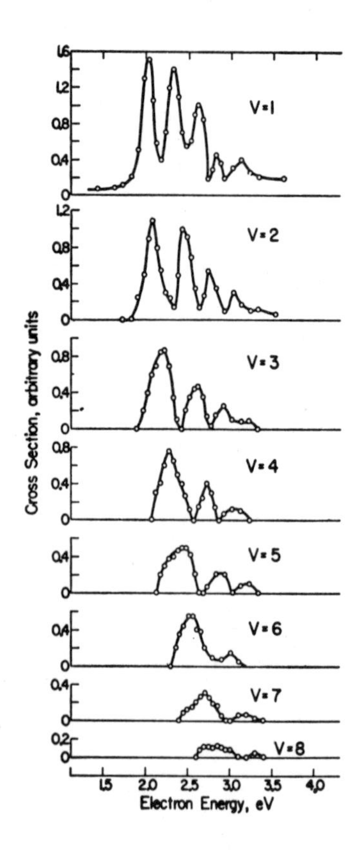

FIG. 4. Cross sections for vibrational excitation of N_2 showing resonance structure. Cross sections are in units of approximately $10^{-16}cm^2$ (12).

vibrational excitation of N_2 as a function of electron energy(12). These results were first presented by Schulz at the Gaseous Electronics Conference in 1961 and were the ones that introduced us to the concept of resonances in molecules. This work showed that the vibrational excitation cross section consists of resonances centered around 2 eV. The cross sections shown are approximately in units of 10^{-16} cm^2 and are very large compared to values expected for non-resonant excitation processes.

In order to calculate the rates of vibrational excitation appropriate to real laser gas mixtures we need the other relevant cross sections, such as the CO_2 cross sections(10) shown as a function of electron energy in Fig. 5. This set of cross sections attempts to take into account all of the available data for elastic and inelastic collisions in CO_2. The upper curve is a cross section for momentum transfer collisions and the lower curves are various vibrational and electronic excitation cross sections, e.g., excitation of the bending modes, of the symmetric stretch modes, and excitation of the 001 level of the asymmetric stretch mode. We notice that the sum of the cross sections for the levels making up the mixed mode is essentially

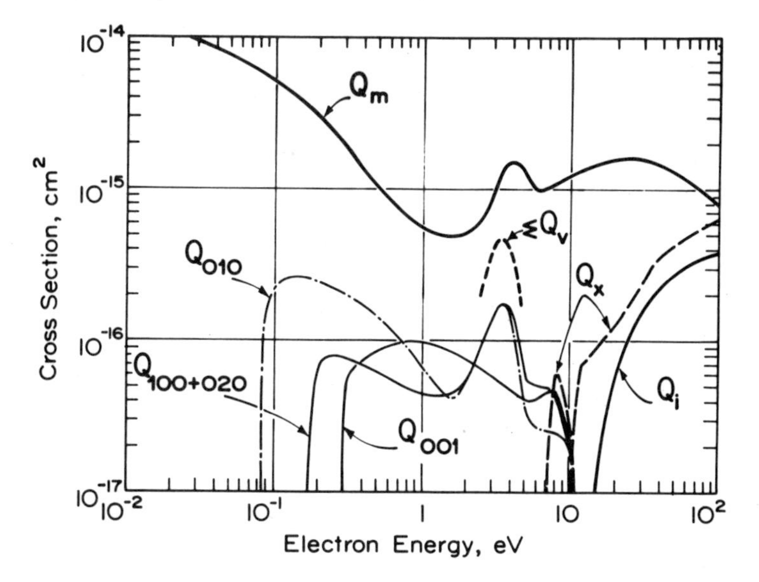

FIG. 5. Cross sections for elastic and inelastic collisions of electrons with CO_2 (10).

equal to or greater than the cross section for excitation of the upper laser level. This unfavorable ratio of cross sections accounts in part for the difficulty in obtaining laser action in pure CO_2, although one must also take into account differences in the relaxation rates of these various levels. Note that there is a large resonance in CO_2, at around 3.8 eV. It turns out that most of this excitation is to the undesirable vibrational levels as far as laser action is concerned. Now that one has sets of cross sections one can calculate electron energy distribution functions, excitation and ionization rate coefficients, electron drift velocities, etc. for various gas mixtures.

B. Electronic Excitations

Figure 6, which is taken from the work of Nigham(9), shows the result of calculations of the fraction of the electron energy which goes into the various levels of nitrogen and CO_2 for a laser mixture that is 10% CO_2, 10% nitrogen, and 80% helium. These results are plotted as a function of the electric field to gas density ratio, E/N. As the value of E/N is increased from low values to high values, or alternatively from low electron energies to high electron energies, there is a transition from a situation in which most of the energy goes into the vibrational modes to a situation in which the energy goes into electronic excitation and eventually into ionization. Over a significant range of E/N, $10^{-16} V \cdot cm^2 < E/N < 3 \times 10^{-16} V \cdot cm^2$, most of the input energy appears as vibrational excitation of the N_2 and is available for transfer to the 001 level of CO_2.

C. Optimum Mode

There has been considerable discussion on the question of the optimum mode of operation of the CO_2 laser(3,19). To a first approximation the gas discharge doesn't really know whether there is a laser operating or not. Therefore, the models of the discharge developed many years ago are applicable, i.e., the E/N at which the discharge operates is such as to satisfy the conditions of balance between electron production and loss at an electron density consistent with the discharge current. This means that when a laser is operated in a self sustained gas discharge mode, e.g., the positive column of the low pressure He-CO_2-N_2 laser(4), the E/N has to be

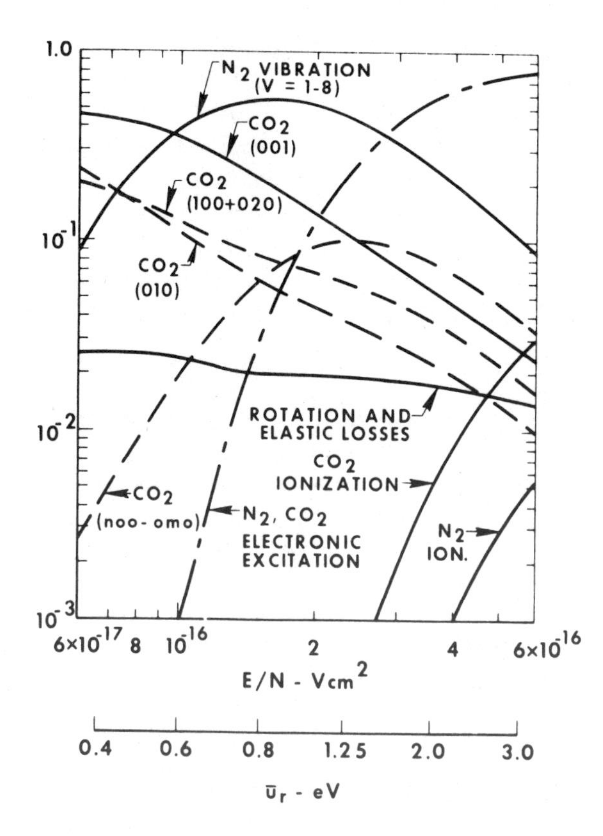

FIG. 6. Calculated efficiencies of electron excitation of various states of CO_2 and N_2 in a 1:1:8 mixture of CO_2, N_2 and He (4).

high enough so that the electrons produce enough ionization to overcome losses due to attachment and electron-positive ion recombination. It has been demonstrated (3,13) that one can maintain the discharge by external ionization, particularly by using high energy electron beams to produce the ionization. In a sense we shouldn't call it a discharge but rather a sort of high power ionization chamber. In this case one can apply almost any electric field to gas density ratio that is convenient. In particular, one can operate at lower E/N values where the transfer of energy to the desired vibrational modes is more efficient and where the growth of instabilities related to the balance between ionization, attachment and recombination is slow.

D. Ionization, Attachment and Recombination

In order to do any detailed analyses of these systems we have to obtain values for the ionization, attachment and recombination coefficients as a function of E/N and gas mixture. Figure 7 shows values of the Townsend ionization coefficient for pure N_2, CO_2 and He as well as some calculations of the Townsend ionization coefficient for some laser mixtures. These results are plotted as a function N/E because such a plot yeilds straight lines which are readily extrapolated to low E/N.

Another coefficient which we need to know something about is the electron-ion recombination coefficient appropriate to the laser mixtures. Of course electron-positive ion recombination has been the subject of a great deal of discussion in these conferences, particu-

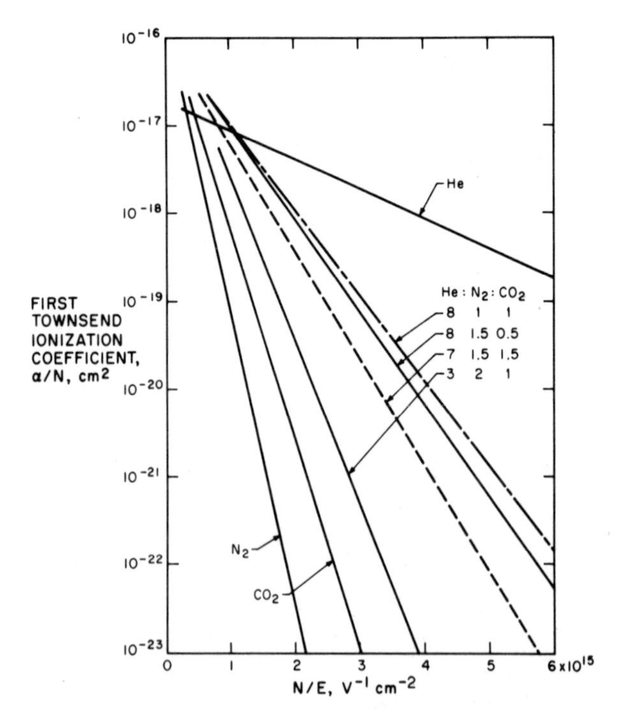

FIG. 7. Calculated Townsend ionization coefficients for various gases and gas mixtures.

larly through the work of Biondi and co-workers(14). Unfortunately, neither theory nor experimental results are available for the mixtures of interest in lasers. In Fig. 8 we have plotted the recombination coefficient for nitrogen as a function of the electron characteristic energy. If the electron energy distribution were Maxwellian this energy would be equal to the temperature expressed in electron volts. The figure shows only a few representative recombination coefficients from the rather large amount of data that is available for the N_2^+ ion(14). As one increases the electron energy one sees a relatively slow decrease with increasing electron energy. Far less is known about complexes such as N_4^+. There is only one experiment in which one has identified the ion, and that is the work of Kasner and Biondi for thermal electrons(15). The dashed curve shows the results of measurement of electron loss by recombination in nitrogen samples typical of those used to prepare laser mixtures(16). The remarkable feature of these data is the very rapid decrease in the apparent recombination coefficient with increasing electron energy. Note that these experiments do not provide any information

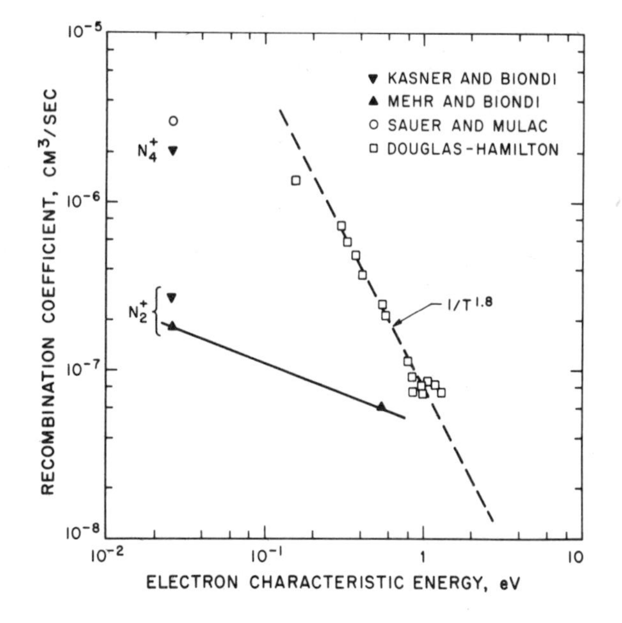

FIG. 8. Electron-positive ion recombination coefficients measured in N_2 by various authors.

as to the identity of the positive ions responsible for the rapidly varying recombination coefficient. On the basis of very recent studies of the effects of impurities on laser discharges by Denes and Lowke(17) one suspects that positive ions formed from water vapor are important in the recombination measurements of Douglas-Hamilton shown in Fig. 8.

E. Low Pressure Discharge

Thus far the greatest degree of success in applying data such as we have been discussing has been with the low pressure CO_2-N_2-He laser.

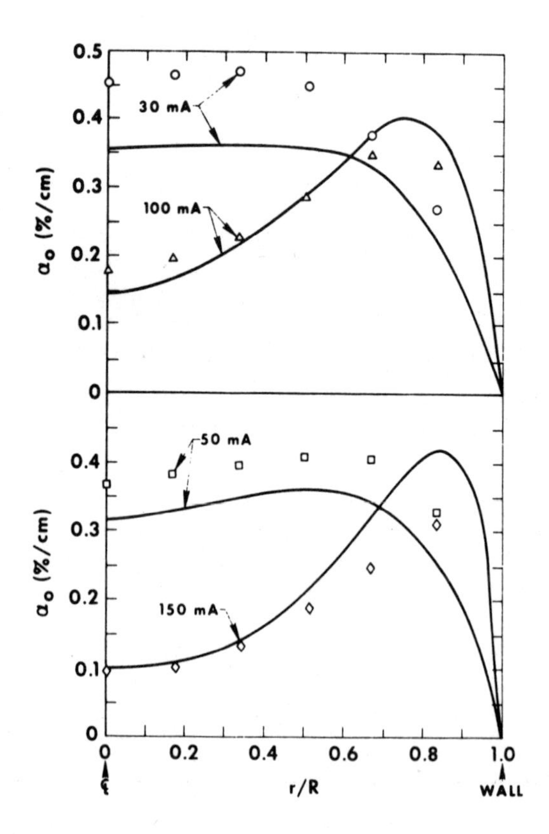

FIG. 9. Measured and calculated gain at 10.6μm vs. radial position for a 1:1:8 mixture of CO_2, N_2 and He operated at 10.5 torr and the currents shown (18).

An example is the results of Wiegan, Fowler and Benda(18) shown in Fig. 9. These authors have calculated the small signal gain as a function of radius for various discharge currents in a typical laser mixture. At low currents they found a maximum gain in the center of the tube and the agreement between the theory and the experiment is within about 15%. As the current is increased gas heating becomes important, the relaxation processes become much more rapid in the center of the tube and the gain goes down. Thus, one sees that in spite of the higher current there is a lower gain in the center of the tube.

F. High Pressure Discharge

Analyses of high pressure CO_2 lasers(13,19) are far less advanced, partly because sufficiently complete experimental data for simple geometries have not been published. This does not stop one from making predictions and comparisons. Thus, the calculated effect of the ionization balance on the efficiency available from various mixtures is shown in Fig. 10, where the predicated excitation effici-

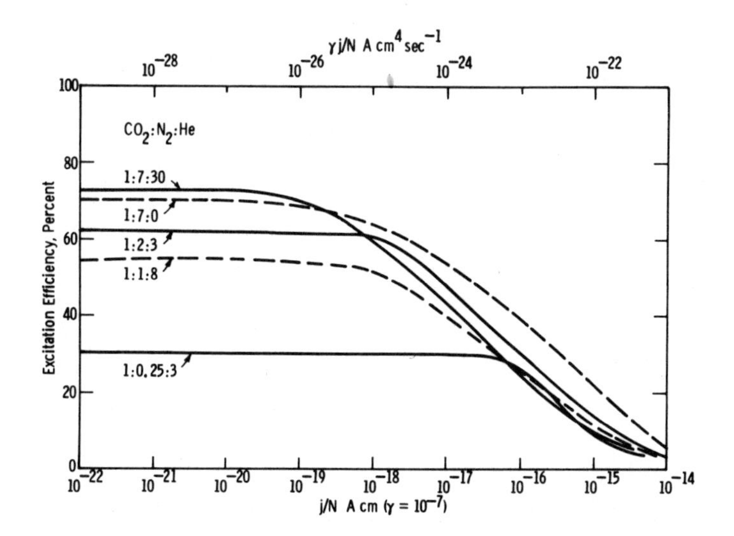

FIG. 10. Calculated efficiencies of excitation of CO_2(001) and N_2 as a function of product of recombination coefficient, current density and reciprocal of gas density (upper scale) and ratio of current density to atom density for a recombination coefficient of 10^{-7} cm^3/sec (lower scale) (10).

encies are plotted as a function of the product of the electron-ion recombination coefficient γ, the discharge current density j and the reciprocal of the gas density. Very recently Denes and Lowke(17) have shown that the effective electron-ion recombination coefficents for the gases used in high pressure commercial lasers is so small ($10^{-8} cm^3/sec$) that E/N of the discharge is independent of j/N for the full range of available current densities. This is equivalent to operation at the left hand side of Fig. 10 and means that the excitation efficiency is fixed and varies from 70% to 30% as the mixture is varied. Note that at the higher $\gamma j/N$ values the excitation efficiency is essentially independent of the mixture used, i.e., the electron energy distribution changes with mixture to yield a roughly constant ratio of vibrational excitation rate to ionization rate. The very low recombination coefficients are consistent with an extrapolation of the recombination coefficient data of Douglas-Hamilton to typical discharge characteristic electron energies of about 2 eV. Note that the theoretical calculations of Fig. 6 also predict that the excitation efficiency decreases as the E/N is raised above the quasiequilibrium values of Fig. 10, as is sometimes done in order to increase the power density which can be transferred to the gas prior to arcing.

V. A LiXe GROUND STATE DISSOCIATION LASER

Now that we have discussed some of the important gaseous electronics processes in some of the relatively well understood lasers I'd like to try to present something that's a bit newer. What can we do about predicting the ability of a proposed system to show efficient laser operation? In other words, the system I'm talking about has not been tried, and maybe it never will be. But let's make a prediction. It is first necessary to discuss a new class of lasers which are sometimes called ground state dissociation lasers because the stimulated emission occurs in a molecule with a dissociating or repulsive ground state potential energy curve (20). Figure 11 shows an example of such a molecule, i.e., the LiXe molecule, as calculated by Baylis (21). To the best of my knowledge no one has ever looked at the spectra of the lithium-xenon molecule, and so we must rely completely on theory. If a molecule is formed in the $^2\Pi$ excited state then that molecule can radiate to the repulsive ground state. Since the ground state is repulsive this molecule flies apart

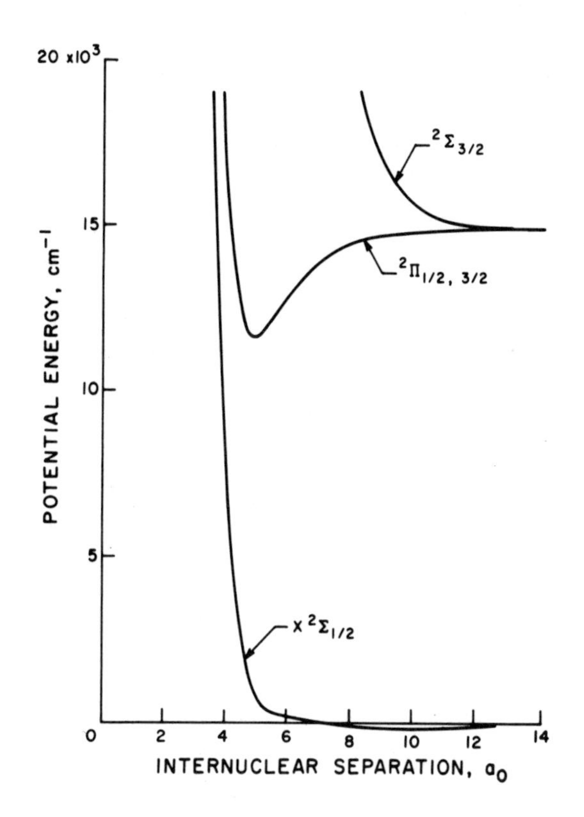

FIG. 11. Calculated potential energy curves for LiXe molecules (21).

in a time short compared to the radiative lifetime of the upper level. If the gas temperature is not too high, the rapid dissociation results in a low ground state population, and we have met one of the conditions necessary for laser operation. The real question is, can one get enough gain to overcome whatever losses there may be in the system? Before we go into that question, let's go back a little bit and ask how we could produce these molecules. The $^2\Sigma$ and $^2\Pi$ potential energy curves shown in Fig. 11 are those formed from an excited lithium atom in one of the ^2P states and a xenon atom. If we produce excited lithium by electron impact excitation from the lithium ground state in the presence of a high density of xenon, there is a possibility of that atom colliding with the xenon atom, of the LiXe molecule being stabilized in a three-body collision, and of the molecule relaxing vibrationally to an equilibrium distribution before the ^2P-^2S resonance radiation escapes to the wall. Recent experiments by Gallagher and co-workers(22) have shown that for similar molecules formed from heavier alkali metals and rare gas atoms the relaxation at pressures of the order of atmospheric pressure is sufficiently fast so it is a good approximation to assume that one has a thermal population of excited molecules. We will make the assumption of an equilibrium vibrational state population in the rest of this discussion. The process just discussed for the production of excited molecules is called an association reaction and has led some people to call the lasers under discussion association lasers.

A. Laser Gain

Let us now try to look a bit more quantitatively at the problem. We assume that what is known as the quasistatic theory of spectral line broadening(22) can be used to calculate the number of LiXe molecules that can radiate at a given frequency. Equations 1 through 4 show the relationships used for prediction of the laser gain (23). Equation 1, taken directly out of Mitchel and Zemansky (24) simply says that the gain g_ν at a frequency ν is proportional to the product of the square of the wavelength, the Einstein A-coefficient and the differences of populations δN of molecules in the upper and lower levels which are capable of stimulated emission or absorption at the photon frequency. The g_u and g_ℓ factors are the statistical weights of the upper and lower states. Thus,

$$g_\nu = \frac{\lambda^2}{8\pi} A_{u\ell} \left[\frac{\delta N_u(\nu)}{\delta \nu} - \frac{g_u}{g_\ell} \frac{\delta N_\ell(\nu)}{\delta \nu} \right] \qquad \ldots\ldots\ldots\ldots\ldots\ldots \quad 1$$

In many molecular problems the Einstein coefficient or radiative transition probability is very much of an unknown. In this particular problem we are dealing with an allowed transition for the atom, and there is theoretical evidence that one should consider the radiative matrix element independent of radius. With this assumption one can calculate the Einstein A-coefficient as a function of radius or frequency.

In the quasistatic theory of spectral line shapes a given frequency ν is a result of the presence of lithium-xenon molecules with an inter-nuclear separation for which the difference in the potential energy curves is $h\nu$. According to Eq. 2 the number of molecules capable of emitting at ν is equal to the product of the number of perturbers within a spherical shell of radius R and thickness δR about the lithium atom in the absence of an interaction, the change in population of perturbers caused by the interaction, and the density of excited lithium atoms. Therefore,

$$\frac{\delta N_{u,j}(\nu)}{\delta\nu} = \frac{4\pi R^2(\nu)\,\delta R(\nu)}{\delta\nu} N_o P_j N_u$$

$$= \frac{4\pi R^2(\nu)}{\left.\frac{\delta\nu}{\delta R}\right|_\nu} \frac{g_{j,u}}{g_{f,u}} [\exp-(\Delta V_u/kT_v)]\; N_o \frac{g_{f,u}}{g_\ell} [\exp-(\Delta V_o/kT_e)] N_\ell \quad \ldots 2$$

In the final form of Eq. 2, we have rewritten the differentials in the form of $\delta\nu/\delta R$, which one might calculate from the difference in upper and lower state potential curves and in terms of the appropriate statistical weights and Boltzmann factors. Here ΔV_u is the change in the potential energy of the $^2\Pi$ states as the internuclear separation R is reduced from infinity, ΔV_o is the excitation energy of the ^2P states of lithium, T_v is the vibrational temperature that describes the relative populations of the vibrational levels of the $^2\Pi$ state, T_e is the excitation temperature used to describe the relative population of lithium atoms in the ^2P excited state of density N_u and in the ^2S ground state of density N_ℓ, k is the Boltzmann constant and the statistical weight ratio $g_{j,u}/g_{f,u}$ takes into account the fact that only a fraction of the ^2P atoms form $^2\Pi$ molecules. If the potential curve for the upper state is repulsive then the energy ΔV_u is positive and there is a reduction in the Xe perturber density. If the potential energy curve is attractive as for the $^2\Pi$ states ΔV_u is negative and there is a buildup of the population of perturbers in the vicinity of the lithium atom. The expression for the gain can be rewritten as

$$\frac{g_\nu}{N_o N_\ell} = \alpha_u(\nu,T_v)\left[1 - \exp[-(\Delta V_\ell - V_u)/kT_v]\exp(\Delta V_o/kT_e)\right] \ldots 3$$

where we have normalized to the ground state density, N_ℓ, and to the perturber density, N_o. The $\alpha(\nu,T_v)$ coefficient contains all terms in the final form of Eq. 2 except the two densities and is the stimulated emission cross section per ground state lithium atom divided by the perturber density. The sign of the bracket then determines whether stimulated emission or absorption is dominant, and leads to the "inversion condition"

$$\frac{\nu_o - \nu}{\nu_o} \geq \frac{T_v}{T_e} \quad \ldots\ldots\ldots\ldots \quad 4$$

The inversion condition says that the ratio of the shift in frequency at which you expect laser operation to the unperturbed frequency ν_o must be larger than the ratio of the vibrational temperature to the excitation temperature for the excited lithium atoms. This condition is relatively easily met in many systems. The real question is: Is the α coefficient in Eq. 3 large enough to give you the gain needed for oscillation?

B. Theoretical Predictions

Figure 12 shows some calculations that we made for the lithium-xenon system using this theory. We've chosen a temperature of 900°K for the vibrational and translational temperature of the gas. We chose 900°K because temperatures of that order are required for achieving

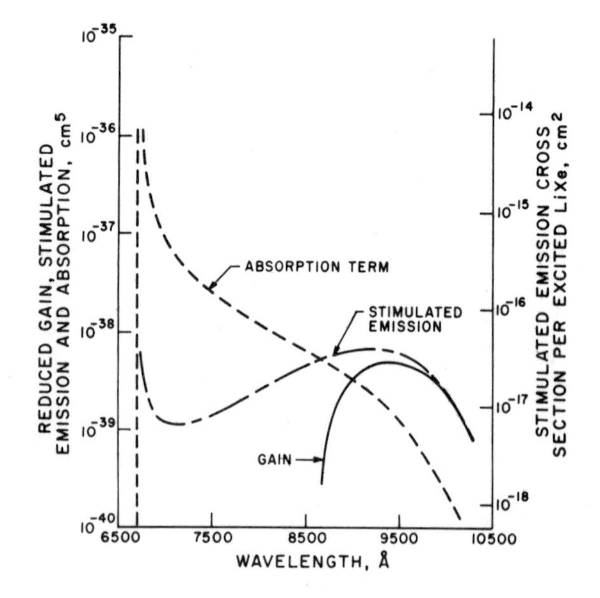

FIG. 12. Predicted stimulated emission, absorption and gain coefficients for LiXe mixture for a gas temperature of 900°K and a Li excitation temperature of 4000°K.

a sufficient lithium vapor density. The assumed excitation temperature, $5{,}000^{o}K$, is fairly typical of what one finds in metal vapor discharge lamps. The curves shown are of the variation with wavelength of the reduced stimulated emission, the absorption and the gain coefficients. These coefficients are to be multiplied by the product of the lithium and xenon densities in order to obtain the gain, etc., per unit length of optical path. The resonance lines of lithium occur at 6710 Å. The absorption term descreases as the wavelength shifts away from the resonance line while the stimulated emission curve goes through a maximum just above 9000 Å. The difference between the stimulated emission and the absorption is the gain. If our potential curves are right, we would expect a peak gain at around 9500 Å, and a gain coefficient of around 3×10^{-39} cm^5. If we assume a xenon density of 10^{20}, i.e., about 10 atmospheres of xenon then the gain cross section is 3×10^{-19} cm^2 per ground state lithium atom. If we assume a density of lithium in the ground state of 3×10^{16} cm^{-3}, then the gain is about 1% per centimeter. These are not unreasonable numbers. Although lithium is a rather difficult vapor to work with, recent developments in heat-pipe technology (25) may help make such experiments feasible. One can possibly get a better feeling for the numbers by expressing the results in terms of gain per excited molecule rather than in terms of the ground state atom density. The scale for the stimulated emission and gain cross sections per excited molecule are shown on the right-hand side of Fig. 12. The peak gain cross section is about 3×10^{-17} cm^2 and the density of excited molecules for our assumed conditions is 3×10^{14} cm^{-3}. This absorption coefficient is roughly what one would expect if one takes the absorption coefficient for an allowed transition and spreads it out over 500 or 1000 Å.

C. Additional Processes

The optimistic picture presented above for a LiXe laser fails to take into account several potentially important processes. Most important is the absorption of the desired photons by the excited LiXe molecules. Since the photon energies of interest are less than the photoionization threshold one needs to know whether there are spectral regions between the lines of the expected absorption bands where the absorption coefficient is less than the stimulated emission coefficient. Such information is not presently available from experiment and its calculation would require a determination of numerous bound and repulsive potential energy curves of the LiXe molecule. Another concern in an electrical discharge is that the electrons with a high mean energy may collisionally excite the vibrational states of the excited LiXe molecule. Other practical considerations include the effect of deexciting collisions between pairs of excited molecules and between excited molecules and Li_2 molecules on the excited state density. Since we are not in a position to predict the importance of these processes, we obviously have a large number of topics for future research.

D. High Pressure Xe Laser

Before leaving the ground state dissociation laser, it should be noted that this principle has been successfully applied to obtain laser operation in high pressure Xe(26). In these experiments a very intense beam of high energy electrons is injected into Xe at several atmospheres pressure and a short pulse of coherent radiation peaking near 1720 Å is observed. Although some information is available regarding the potential energy curves for the Xe_2 molecule, it is not certain which excited states are responsible for the stimulated emission. The presence of a significantly large stimulated emission cross section means that a radiative transition with a reasonably large oscillator strength is involved. However, we have no idea as to whether this is the property of a single excited state or is the result of collisional coupling of metastable and radiating states. It is expected that the considerable current interest in the rare gas molecular lasers will spur theoretical and experimental investigations of these molecules.

VI. CONCLUSION

I hope that these examples serve to demonstrate the close connection between laser technology and the areas of interest of those researchers who participate in the Gaseous Electronic Conferences. We look forward to a continued fruitful exchange between these two areas of science.

ACKNOWLEDGEMENT

This work was supported in part by the Advanced Research Projects Agency of the Department of Defense and was monitored by U.S. Army Research Office - Durham, under Contract No. DAHC04-72-C-0047.

* Staff member, Laboratory Astrophysics Division, NBS and Professor Adjoint, Dept. of Physics and Astrophysics, Univ. of Colorado.

REFERENCES

1. W.R. Bennett, Jr., Applied Optics Supplement on Optical Masers, p. 4 (1962).
2. C.K. Rhodes and A. Szoke, in Laser Handbook edited by F.T. Arecchi and E.O. Schulz-Dubois (North-Holland, Amsterdam, 1972), Vol. 1, Chap. B1.
3. A.J. Demaria, Proc. I.E.E.E. 61, 731 (1973).
4. R.H. Bullis, W.L. Nigham, M.C. Fowler and W.J. Wiegand, AIAA J. 10, 407 (1972).
5. A. Javan, W.R. Bennett, Jr., and D.R. Herriott, Phys. Rev. Letters 6, 106 (1961).
6. H.S.W. Massey, Electronic and Ionic Impact Phenomena (Oxford University Press, London, 1971) Vol. 3, Chap. 18.
7. W.T. Silfvast, Appl. Phys. Letters 13, 169 (1968) and 15, 23 (1969).
8. C.E. Webb, A.R. Turner-Smith and J.M. Green, J. Phys. B: Atom. Molec. Phys. 3, L135 (1970).
9. W.L. Nigham, Phys. Rev. A 2, 1989 (1970).
10. J.J. Lowke, A.V. Phelps and B.W. Irwin, J.Appl. Phys. 44 (1973)
11. R.L. Taylor and S. Bitterman, Rev. Mod. Phys. 41, 26 (1969).
12. G.J. Schulz, Phys. Rev. 125, 229 (1962) and 135, A988 (1964).
13. C.A. Fenstermacher, M.J. Nutter, W.T. Leland and K. Boyer, Appl. Phys. Letters 20, 56 (1972).
14. See for example J.N. Bardsley and M.A. Biondi in Advances in Atomic and Molecular Physics, edited by D.R. Bates and I. Esterman (Academic Press, New York, 1970), Vol. 6.
15. W.H. Kasner and M.A. Biondi, Phys. Rev. 137, A317 (1965).
16. D.H. Douglas-Hamilton, J.Chem. Phys. 58, 4820 (1973).
17. L.J. Denes and J.J. Lowke, Appl. Phys. Letters 23 (August) (1973).
18. W.J. Wiegand, M.C. Fowler and J.A. Benda, Appl. Phys. Letters 18, 265 (1971).
19. G.C. Vlases and W.M. Moeny, J.Appl. Phys. 43, 1840 (1972); S.J. Kast and C. Cason, J.Appl. Phys. 44, 1631 (1973).
20. F.G. Houtermans, Helv. Phys. Acta. 33, 933 (1960).
21. W.E. Baylis, J.Chem. Phys. 51, 2665 (1969).

22. R.E.M. Hedges, D.L. Drummond and A. Gallagher, Phys. Rev. A 6, 1519 (1972).
23. A.V. Phelps, Tunable Gas Lasers Utilizing Ground State Dissociation, JILA Report No. 110, September 1972.
24. A.C.G. Mitchell and M.W. Zemansky, Resonance Radiation and Excited Atoms (University Press, Cambridge 1934), Chap. III.
25. C.R. Vidal and F.B. Haller, Rev. Sci. Instr. 42, 1779 (1971).
26. H.A. Koehler, et al., Appl. Phys. Letters 21, 198 (1972); J.B. Gerado and A.W. Johnson, IEEE J. Quantum Electron. QE-9, 748 (1973).

Gaseous Electronics, eds. J.Wm. McGowan and P.K. John

GASEOUS ELECTRONICS IN ELECTRIC DISCHARGE LAMPS

J.F. WAYMOUTH

GTE Sylvania Incorporated
Lighting Products Group
Danvers, Massachusetts

I. INTRODUCTION

The discharge lamp industry is based on electric discharges in gases, with electron temperature 0.5-1.0 ev; because these in general combine high emissivity in spectral regions of desired output and low emissivity everywhere else, together with the maximum of the black body curve in the desired spectral region, they offer high efficiency of conversion of electrical power into light. Since there is a great variety of such discharges and corresponding discharge lamps, it will be necessary for me to select a rather limited number of topics for discussion today. Consequently, I will make no attempt to make my discussion all inclusive, but will try instead to highlight a few examples of phenomena common in Gaseous Electronics that are important in discharge lamps, and a few examples of the application of well-known plasma measurement techniques to the analysis of phenomena in discharge lamps. I will limit myself to two families of lamps from which to choose my examples: 1) The low-pressure mercury-rare-gas discharge used in fluorescent lamps, which can now be said to be quite well understood; and 2) The high-pressure metal halide discharge lamp, which cannot.

It is unfortunately true that the necessity of limiting the topics for discussion prevents consideration of many areas where outstanding work has been done; to those workers whose contributions will not be discussed, I apologize in advance.

II. THE POSITIVE COLUMN OF FLUORESCENT LAMPS

Fig. 1 illustrates the configuration and some of the important

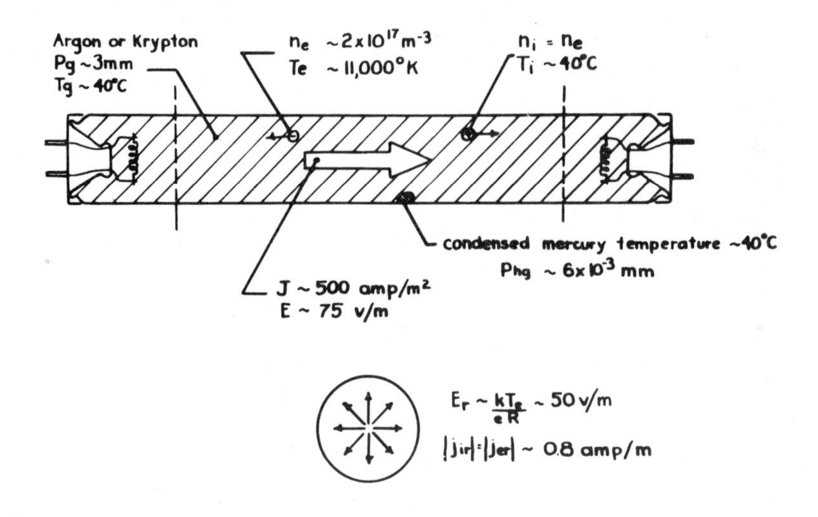

FIG. 1. Some parameters of a 40 Watt fluorescent lamp.

parameters of the low-pressure mercury-rare-gas discharge in a fluorescent lamp. This is a diffusion-controlled discharge in a rare gas at a few torr, plus saturated mercury vapor at a few millitorr. Fully 60 percent of the electrical energy dissipated in the positive column is radiated in the 2537Å resonance line of mercury. Approximately 25 percent of the energy is dissipated in elastic collisions between electrons and rare gas atoms.

The rare gas serves to control the electron temperature by retarding ambipolar diffusion and to reduce the loss of high energy electrons to the walls which would occur if mercury at 6 millitorr were the only filling gas. The lamp is actually more efficient with rare gas than without it, despite the increased elastic collision loss.

Of the remaining 15 percent of the input energy, about one-half is radiated in the 1850Å resonance line and the balance is distributed among the other lines of mercury. Ionization loss is about one percent, and there is no excitation of rare gas lines.

Langmuir probe measurements have been used by Easley (1) and by Verweij (2) to determine electron density and temperature in such discharges. A sample of Verweij's results is shown in Fig. 2.

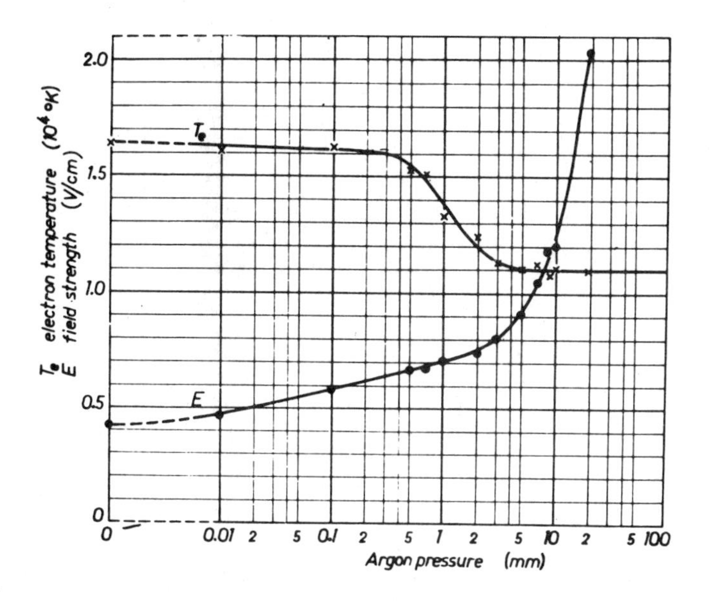

FIG. 2. Electron temperature and electric field strength in a mercury-argon discharge at 400 ma current, 6 millitorr Hg Pressure, 36 mm ID tubing, as a function of argon pressure (after Verweij)(2).

Extensive measurements of radiation output have been performed by a number of people, of which the recent ones of Koedam, Kruithof and Riemens (3) appear to be the best. Fig. 3 illustrates a portion of their results.

Carl Kenty, well known to most of us here, carried out the first attemps to quantitatively analyze(4) such a discharge, calculating rates of excitation, de-excitation and radiation from measured values of electron density and temperature and estimates of cross-sections.

Fig. 4 shows Kenty's principal results, comparing calculated excitation rates with measured radiation outputs(5) for the various transitions. Kenty also used absorption measurements to determine concentration of excited 3P atoms. The degree of balance achieved for the several states was sufficiently good to endow the estimated cross-sections with a considerable degree of confidence.

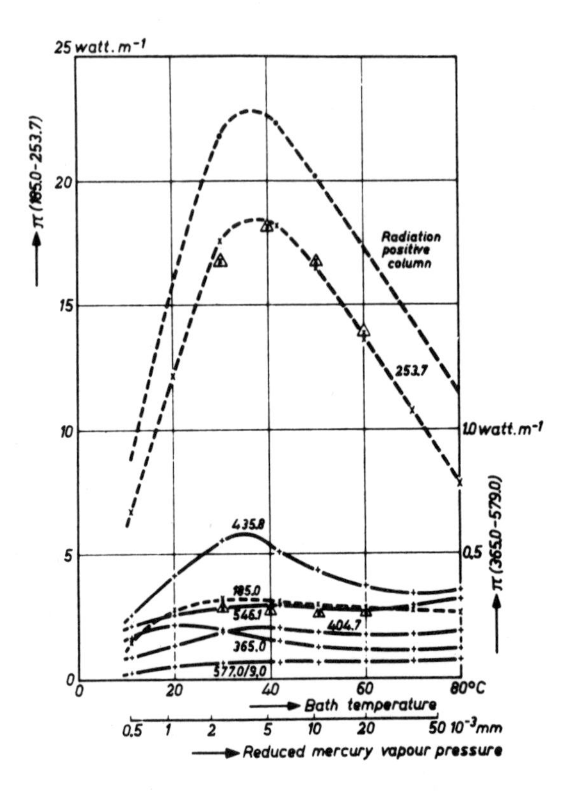

FIG. 3. Radiation output of various lines of mercury spectrum in a mercury-argon discharge at 3 torr argon pressure, 400 ma discharge current, 36 mm ID tube, as a function of mercury vapor pressure (after Koedam, Kruithof and Riemens)(3).

The problem of cross-sections is still a major difficulty; even to-day, more than twenty years later, Kieffer's compilation (6) lists excitation functions for 34 lines of mercury which do not include 2537Å or 1850Å, and all of which deal only with excitation from the ground state. Note however, that the majority of the excitation of the upper state comes stepwise via the triplet-P states.

Fig. 5 shows the cross-sections used by Kenty(4), taken from Yavorsky's quantum mechanical calculations(7), for excitation of 7^3S_1 (the upper state for the 5461Å, 4358Å, and 4047Å lines). It is clear from these cross-sections that the majority of the excitation of 7^3S_1 will come as a result of excitation of the ^3P states. So

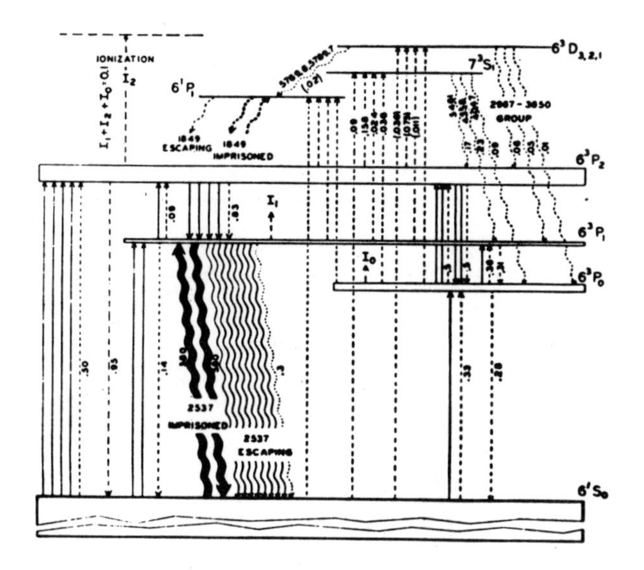

FIG. 4. Excited state balance, mercury-argon discharge, according to Kenty (4).

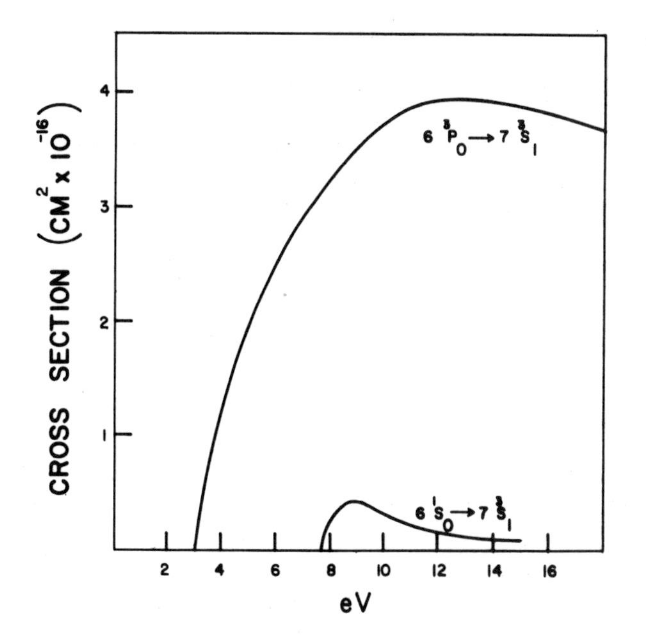

FIG. 5. Cross-sections for excitation of 7^3S, excited state of mercury from the ground state (6^oS_o), and from the lower metastable (6^3P_o), according to Yavorsky(7) as cited by Kenty(4).

far as I am aware, no one has experimentally measured cross-sections for second excitation of excited states of mercury.

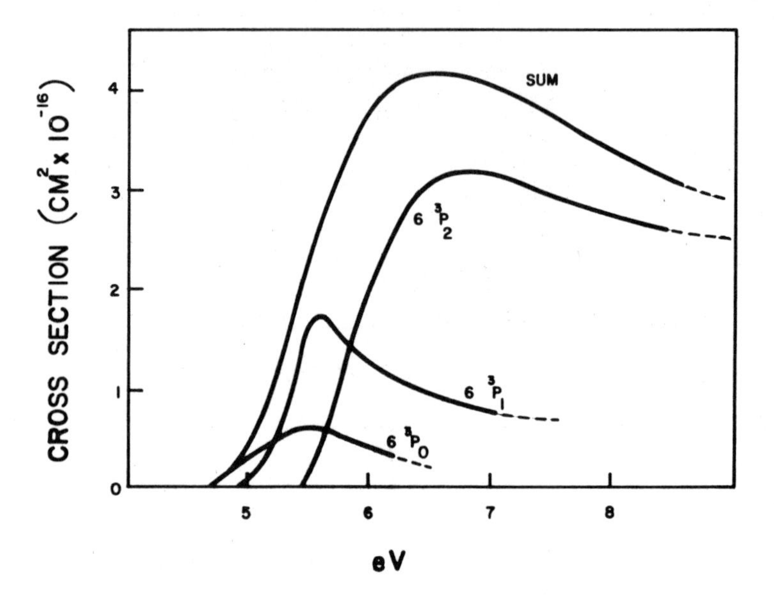

FIG. 6. Cross-sections for excitation of triplet-P states of mercury from the ground state, according to Kenty(4).

Fig. 6 shows the cross-sections used by Kenty for the excitation of the ^3P states from the ground state, arrived at by a blend of experimental data, quantum mechanical calculations and judicious guesswork(4).

Subsequently, these calculations were extended by myself(8), and later by Cayless(9), combining excitation rate energy loss calculations using Kenty's cross-sections with Schottky-Von Engel-Steenbeck ambipolar diffusion theory. Herein lies another problem; most of the ionization in this discharge is stepwise via the ^3P states, but no cross-section for this process is known. I used it as an arbitrary parameter adjusted to fit calculated electron temperature to that measured by Easley (1) at one particular value of dishcarge current and mercury vapor pressure.

Fig. 7 shows calculated T_e and n_e vs mercury vapor pressure. Normalization point for T_e is at $40^{o}C$; no adjustment is made in the electron density values. Normalization of electron temperature yields a value of ionization cross-section of ^3P states 1.5 x that of the ground state. If the ^3P states were in thermal equilibrium with the electron gas, and if the electron energy distribution were Maxwellian to well above the ionization potential, this would lead to 13.5 x as much ionization via the stepwise process as via the direct. Actually the ^3P states range between 10 and 30% of LTE, but the electron energy distribution falls well below Maxwellian for energies above 8 to 9 Volts.

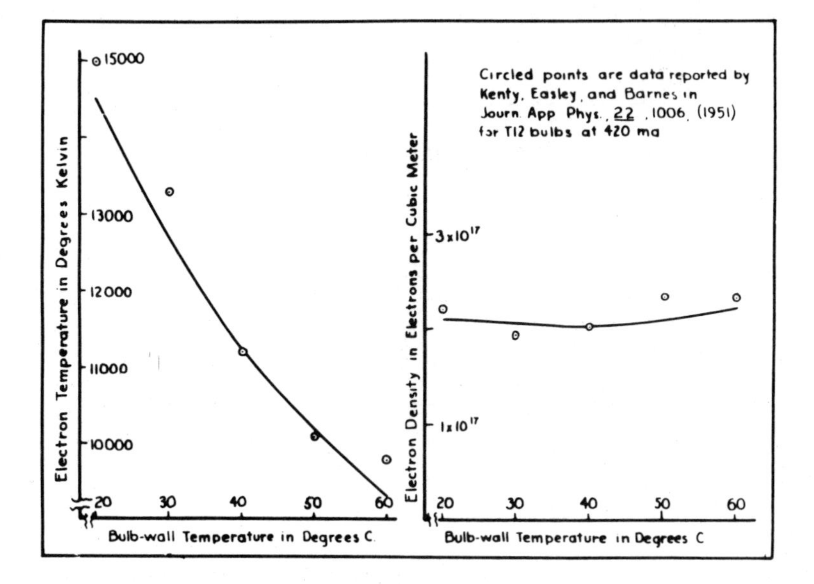

FIG. 7. Comparison of calculated and experimental electron temperature and density in a mercury-argon discharge at 420 ma in a 1.5 inch OD bulb (36 mm ID) as a function of condensed-mercury temperature (8).

For ease of comparison with engineering results, I have converted calculated ultraviolet output into lumens by including a conversion factor for a so-called "Cool White" phosphor.

Fig. 8 plots actual lumens vs calculated lumens at rated current for

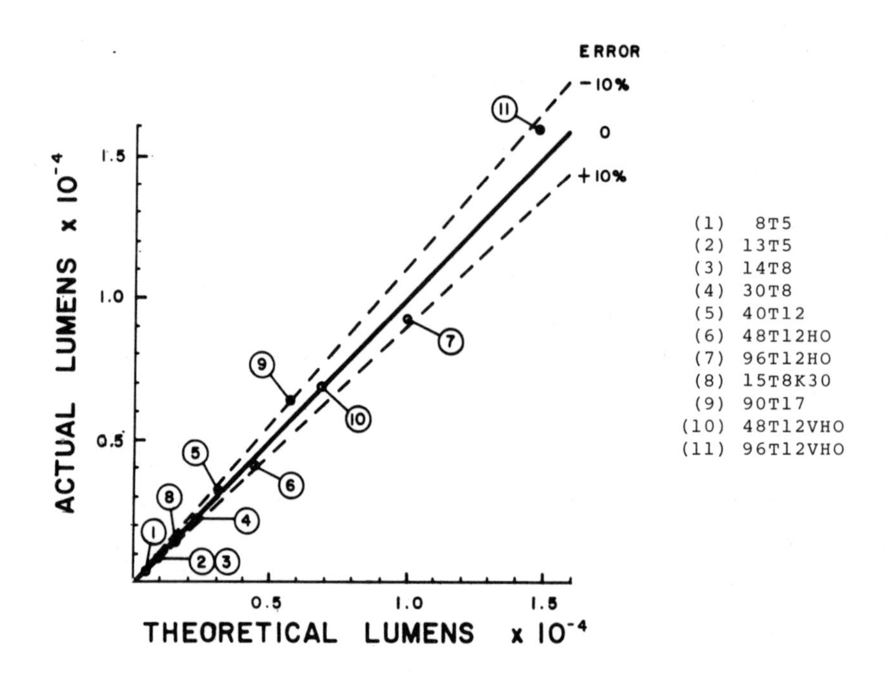

FIG. 8. Comparison of actual lumens vs calculated lumens at rated current for eleven fluorescent lamps of various diameters, lengths, rare gas fills, and power consumption. Numbers refer to lamps.

eleven different fluorescent lamps ranging in diameter from 1.5 cm to 5 cm, length from 10 cm to 2.5 meters and in power consumption from 8 Watts to 200. Neon, argon, and krypton filled lamps are included in the sample.

There is similar agreement between actual power vs calculated power at rated current for the same lamps and the actual arc drop vs calculated arc drop at rated current. The agreement between calculated and measured quantities (within ± 10%) is good enough that we can use this model as a guide to the design of new fluorescent lamps.

III. CATHODE PHENOMENA IN FLUORESCENT LAMPS

The cathode mechanism in a fluorescent lamp is a hot-cathode thermionic arc, with cathode fall of the order of magnitude of the ionization potential of mercury. One of the technological triumphs of the industry has been the development of oxide type cathodes which will survive as much as 30,000 hours of continuous operation, and several thousand "starts" consisting of the simultaneous application of cathode heater voltage and open circuit voltage to the device. I would like now to discuss the behavior under operating conditions of such cathodes, which consist of multiply-coiled baskets of tungsten wire impregnated with alkaline earth oxides. Thermionic emission takes place from the exposed surface of activated oxides between the turns of tungsten wire, enhanced by accelerating fields provided by the ion space charge in the cathode sheath.

The following equations govern the interaction of such a cathode with the discharge:

$$I = I_+ + I_-$$

$$\left.\begin{aligned} I_- &= I_o(T_k)\ M(E,T_k) \\ T_k &= f(P_h, I^2R, V_k,\ I\Phi_c) \end{aligned}\right\}\ \text{Cathode}$$

$$E_k = \phi(I_+,\ V_k)\ :\ \text{Sheath}$$

$$I_+ = \varepsilon(V_k) I_-\ :\ \text{Negative Glow}$$

I is total current, I_+ is total ion current to the cathode, and I_- is total electron emission from the cathode. $I_o(T_k)$ is the zero-field thermionic emission current from the cathode at temperature T_k while $M(E,\ T_k)$ is the multiplication of this zero field emission by the anomalous Schottky effect at the cathode surface field strength E and temperature T_k. The cathode temperature is a function of the net power delivered to it, including heater power P_h, I^2R heating in the cathode resistance R, ion bombardment heating proportional to cathode fall V_k, and electron emission colling proportional to calorimetric work function ϕ_c. The cathode surface field strength E_k is a function of ion current and cathode fall, V_k, while ion current is a function of cathode fall, V_k, and electron current. Many of these variables can be measured, and the rest calculated from these equations, giving a reasonably self-consistent picture of the cathode behavior. Consider first of all the thermionic emission at zero field.

Fig. 9 qualitatively shows potential distribution in the discharge in front of the cathode for three different conditions: A, zero field thermionic emission current less than discharge current; B, equal to discharge current; C, greater than discharge current. In case C, there must be a retarding field at the cathode, and a potential minimum in front of the cathode, exactly as in the case of a vacuum tube under space-charge-limited conditions. Positive ions, generated in the positive column and drifting toward the cathode, can get trapped in this potential minimum and oscillate. These ion oscillations can be detected external to the tube. To measure zero field thermionic emission, one need only fix the cathode temperature by external heat, and slowly increase a D.C. discharge current to the point where RF noise due to ion oscillation stops.

A variety of other techniques may be employed for this measurement

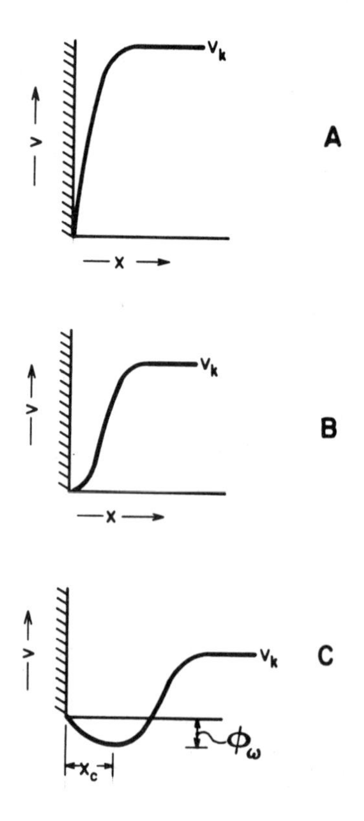

FIG. 9. Potential distribution from cathode through sheath to negative glow for zero field thermionic emission, A, less than, B, equal to, C, greater than, discharge current(10).

such as have been described by Found(11), Druyvestyn and Warmoltz (12), Forster-Brown and Cayless(13) and by Speros and Bucilli(14). The RF noise technique correlates well with these, and has to my mind advantages of simplicity, not requiring special lamps, and being very "gentle" - it does not disturb the state of activation of the cathode.

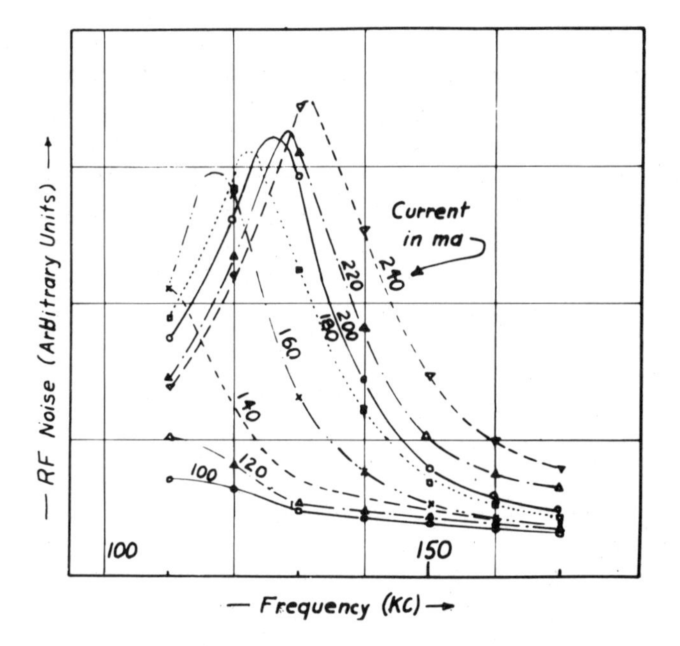

FIG. 10. Frequency distribution of RF noise for various discharge currents, for a cathode with a zero field thermionic emission of 320 ma. This type of distribution and the shift to higher modal frequencies with increasing current is consistent with a model of ions oscillating in a potential well produced by an electron space charge (10).

Fig. 10 shows the frequency distribution of RF noise for various currents for a cathode with a zero field thermionic emission of 250 ma. This behavior of noise-vs-frequency is consistent with the concept that mercury ions are oscillating in a potential well; the shift in frequency as current increases is consistent with the diminishing thickness and depth of the potential minimum as discharge current approaches zero field emission current(10).

For most cathodes, maximum zero-field thermionic emission is typically 10-20 percent of the steady-state arc current of the cathode at the same temperature. Therefore our attention turns now to phenomena in the negative glow. The cathode operates in an accelerating electric field due to the ion space charge in a cathode sheath, across which the cathode fall of potential appears. The accelerating field increases the thermionic emission due to the anomalous Schottky effect.

The magnitude of the accelerating field is proportional to the square root of the ion current density, and hence the total electron emission is strongly levered by the ion current. The ions themselves are produced in the negative glow, a bright blob of plasma surrounding the cathode, by a flux of "primary" electrons accelerated by the cathode fall into the negative glow, where they are elastically scattered by rare gas atoms but essentially retain their identity as primary electrons until they make an inelastic collision creating either an excited atom or an ion. The ions diffuse by ambipolar diffusion in the weak electric fields in the negative glow, to the tube walls, cathode supports, as well as to the surface of the cathode sheath. Only a fraction of the ions created in the negative glow reach the cathode.

These notions derive from a series of Langmuir probe investigations of the negative glow carried out by myself. Measurements were made with a series of tubes containing movable plane probes, using a pulsed-probe technique(15) capable of maintaining a clean surface to within a millimeter of a hot oxide-cathode, no mean feat.

In the center of the negative glow, Maxwellian distributions are found. Data at 3 and 4.25 cm are obtained in the Faraday dark space, and show a deficiency of high energy electrons. Data near the cathode show a strong contribution due to primary electrons.

Fig. 11 illustrates the electron energy distributions calculated from such data via the Medicus modification(16) of Druyvestyn's method, and demonstrates a high energy group, the primary electrons, which have energies approaching the cathode fall but are randomly oriented in direction.

Penetration of the primary electrons into the negative glow can then be measured. Fig. 12 gives the experimental value of penetration coefficient α vs a theoretical one calculated on the assumption that primary electrons diffuse away from the cathode but retain their energy and their identity until they make an inelastic collision (17). The calculated lifetime for this process is about an order of magnitude shorter than the thermalization time.

The important consequence of this is that, by suitable choice of rare gas and mercury vapor pressure, it is possible to make the penetration distance of primary electrons, and hence the distance from the cathode at which the average ion is produced, as much as one centimeter.

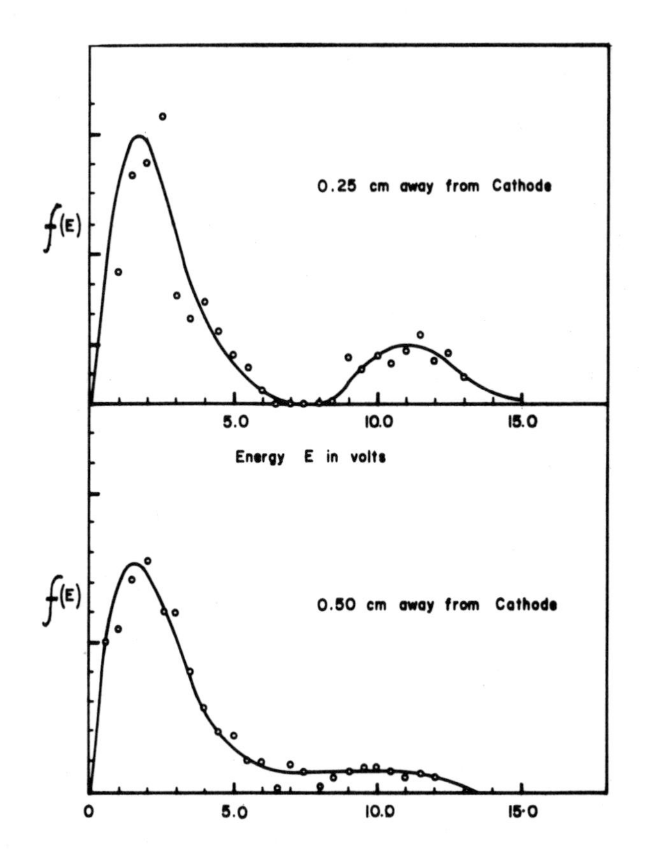

FIG. 11. Electron energy distributions derived from data taken from Ref. 15. Note high energy group, the primary electrons, having energy approximately equal to cathode fall.

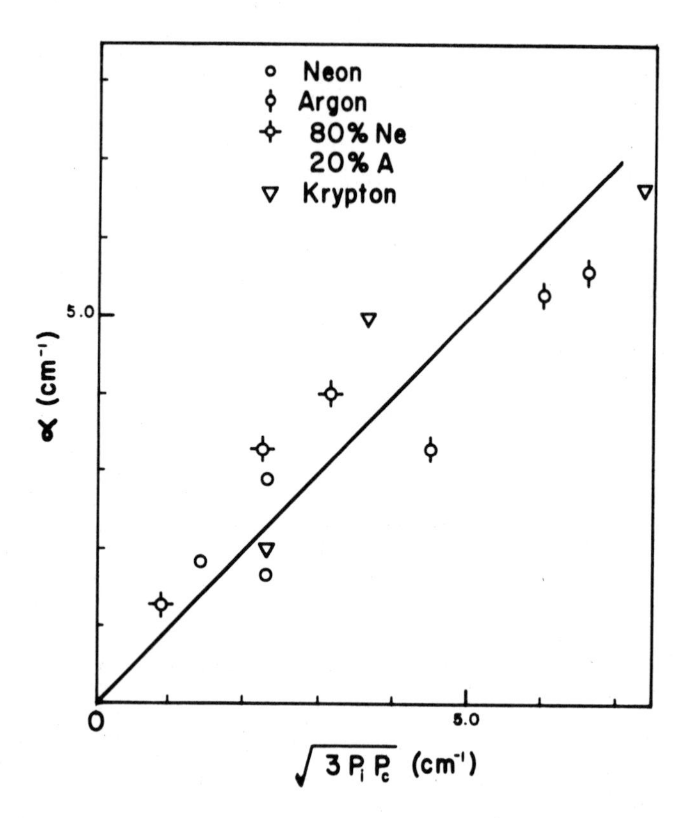

FIG. 12. Experimental penetration coefficient for primary electrons vs. penetration coefficient calculated assuming that primary electrons make elastic collisions randomizing their direction of motion but substantially retain their energy (and hence their identity in the electron energy distribution) until they make inelastic collisions (15).

This permits the use of the geometry of Fig. 13 for measuring ion current to a cathode (18). Two coils of tungsten wire, one bare, the other coated with emission oxides, are interwound but isolated from each other. The cathode sheath thickness, 0.1 to .03 mm, is small in comparison to spacing between the coils, while the average ion is produced at a distance large in comparison to the spacing, and has an approximately equal chance of reaching either. The coated coil emits electrons, while the bare coil is operated at a sufficiently high temperature and negative bias that it is maintained free of oxide contaminants and its thermionic emission is that of clean tungsten, essentially negligible at that temperature.

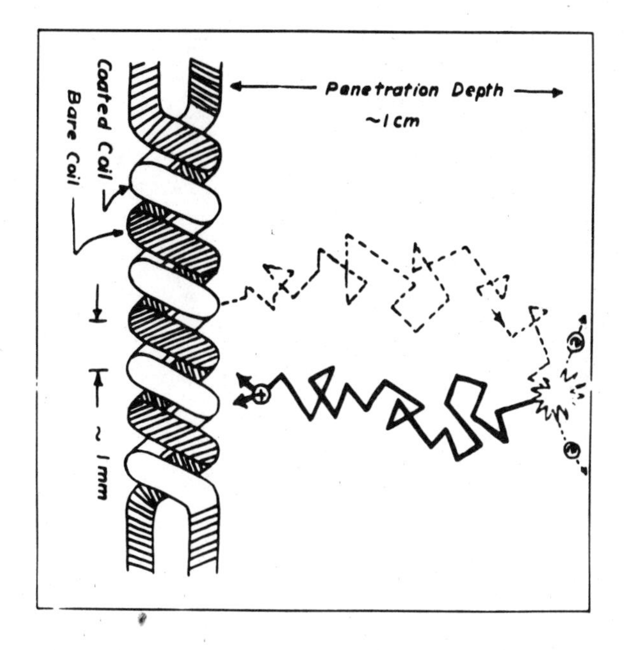

FIG. 13. Design of cathode structure for measuring ion current to cathode(18).

Thus, the current to the bare coil is only ion current, which is equal to the ion current to the coated coil, while the current to the coated coil is electron emission plus ion current. The ion-to-electron current ratio of the total cathode structure is twice the current to the bare coil divided by the current to the coated coil minus the current to the bare coil.

A plot of this ratio vs cathode fall at 1.0 torr 80% Ne, 20% Ar plus 1 milli-torr of mercury is shown in Fig. 14 as a function of cathode fall. Two-stage ionization of mercury, dependent on total current density, is evident for low values of cathode fall, while for cathode fall above 11-12 volts the principal process is believed to be Penning ionization: Creation of argon metastables and their subsequent ionization of mercury.

It would be quite cumbersome to try to construct large numbers of interwound cathodes for study of cathode coating behavior as a function of ion current. Accordingly, use is made of the fact that,

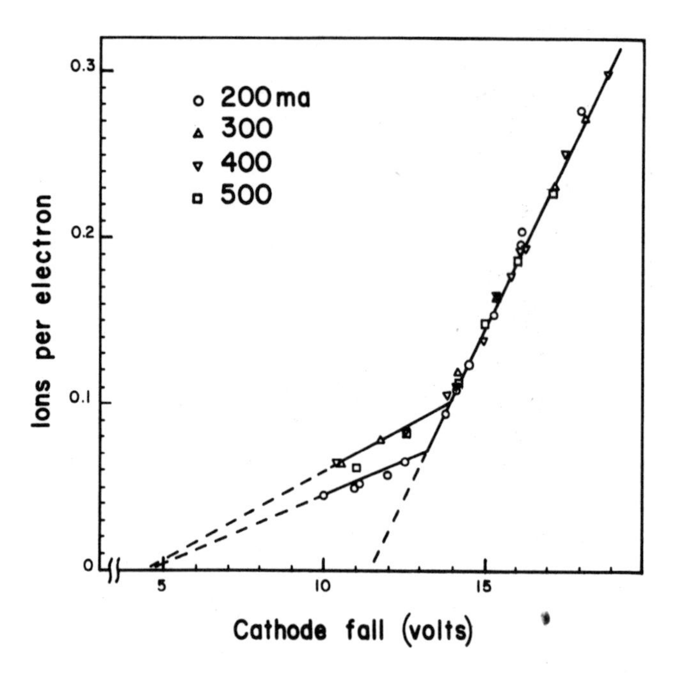

FIG. 14. Ion-to-electron current ratio as a function of cathode fall in a mercury-rare-gas discharge at 1 millitorr Hg pressure, 1 torr 80% Ne-20% A (18).

provided test cathode dimensions, support dimensions and position, fill gas and mercury pressure and tube diameter are the same as those here, the boundary conditions for the ambipolar diffusion problem are the same, and the same fraction of total ions generated will reach the cathode. Therefore we need only measure cathode fall for the test cathode and use these data to calculate I_+/I_-.

Measurements of cathode temperature, ion current, total electron current, and zero field thermionic emission, of a large number of cathodes demonstrated that various possible secondary electron emission processes were incapable of supplying more than a few percent of the electron emission and that the principal process was multiplication of the zero field thermionic emission by the anomalous Schottky effect due to the several thousand Volts/cm field at the cathode surface.

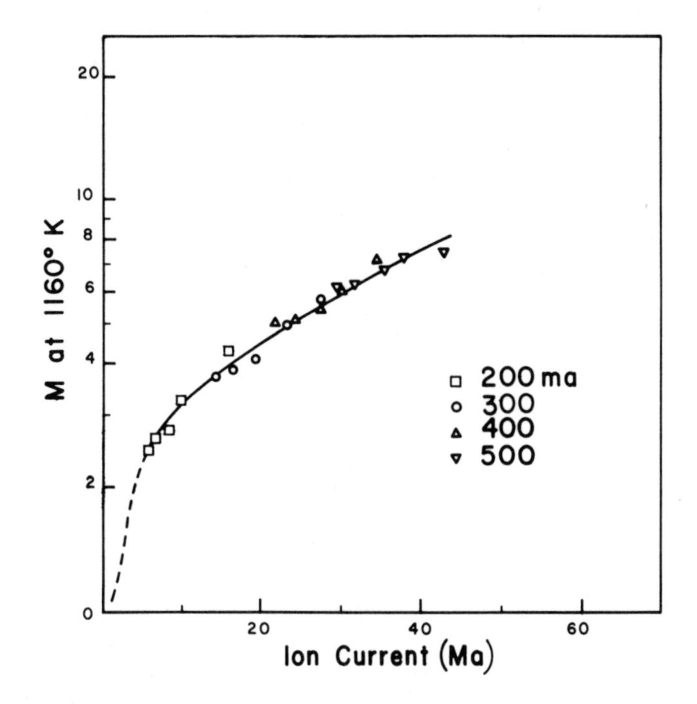

FIG. 15. "Multiplication" of zero-field thermionic emission by the anomalous Schottky Effect, reduced to 1160°K cathode temperature as a function of ion current to the cathode(18).

Fig. 15 illustrates the multiplication factor as a function of ion current (proportional to E_k^2) for a typical cathode. These results are felt to be quite reasonable for a heterogeneous patchy emitter such as these cathodes, and the value of approximately 1 volt for the difference in work function between high and low work function patches which is obtained from such data is quite consistent with other oxide cathode knowledge.

IV. METAL HALIDE LAMPS

The final area I would like to discuss today deals with a phenomenon of importance in metal-halide high pressure lamps. These are high-pressure arcs in quartz tubes containing mercury and one or more metal iodides (19).

FIG. 16. Photograph of a metal halide arc lamp.

Fig. 16 is a photograph of a typical lamp. Such lamps combine high efficiency, good spectral balance, and high light output, and are increasingly displacing fluorescent lamps in a number of commercial installations.

A phenomenon which I call "radiative constriction" of the arc was encountered in the development of one type of such lamps, and the overcoming of it was necessary for its successful introduction to the marketplace. This phenomenon can occur in any high pressure discharge in local thermal equilibrium in which the electrical conductivity has a steeper temperature coefficient than the radiation source function. This is given by the following equations (20) governing the radial temperature distribution in a thermal equilibrium arc:

$$\nabla\ (-H\nabla T) = P_H$$

$$P_H = P_e - P_r + P_a$$

$$P_e = j \cdot E_z$$

$$= en_e \mu E_z^2$$

$$n_e = n_o \frac{\frac{1}{2}(2\pi mkT)^{\frac{3}{4}}}{h^{\frac{3}{2}}} \exp(-eV_i/2kT)$$

$$P_r = An_o \; h\nu \; \exp(-e\overline{V}/kT)$$

$$P_a = n_o h\nu\sigma\Gamma$$

The radial temperature profile is determined by the conditions that the divergence of the heat flux is equal to the local production of heat, which can be seen to be equal to the difference between local

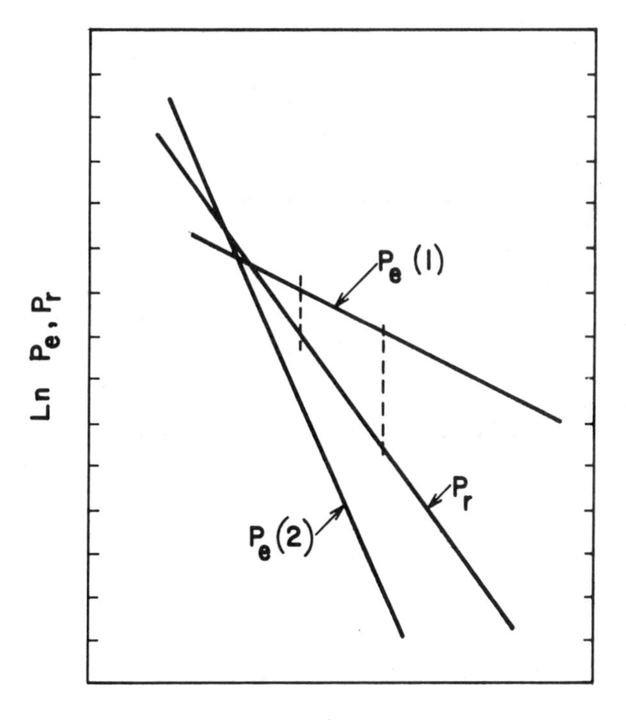

FIG. 17. Illustrating two possible temperature-dependences of local electrical power input for a given temperature dependence of the local radiation loss. See text for discussion (20).

electrical power input and local radiation loss. The electrical power input is proportional to the conductivity which varies as exp $(-eV_i/2kT)$. In an arc in which radiation is mainly in optically thin lines, radiation output can be expressed as approximately proportional to exp $(-e\overline{V}/kT)$ where $\overline{V}$ is an average excitation potential. Clearly, at $R = 0$, $P_h = P_e - P_r$ must be greater than zero so that curvature of temperature is negative at the arc axis.

Fig. 17 illustrates two possible situations for a given radiation source function corresponding to an average excitation potential $\overline{V}$. First $V_i/2 < \overline{V}$; electron density, and hence local electrical power

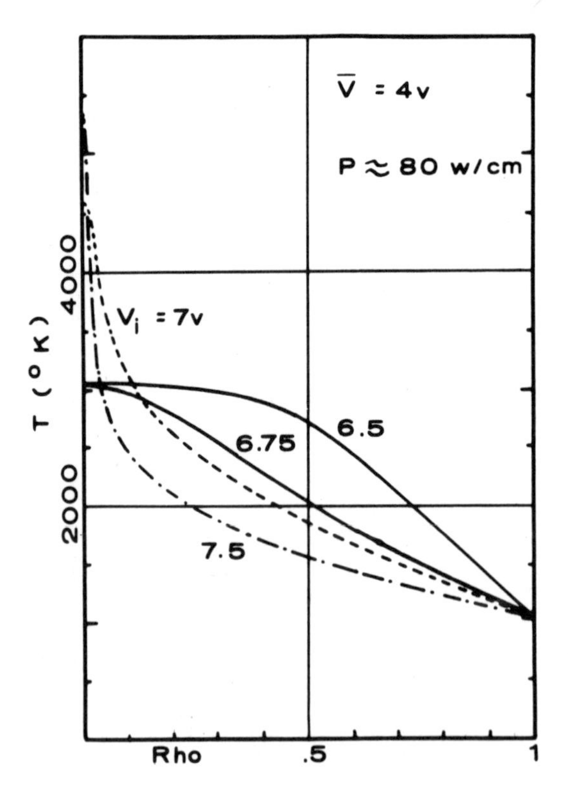

FIG. 18. Calculated radial temperature dependence for an arc in mercury at 2500 torr, plus a metal at 25 torr having an average excitation potential of 4 volts, for various assumed values of ionization potential. Note the transition from wall-stabilized to constricted temperature profiles for ionization potentials greater than 6.75 V.

input decrease more slowly with T than P_r, $P_e > P_r$ for all lower temperatures, and $\nabla^2 T$ is always negative. This results in wall-stabilized temperature profiles.

If, on the other hand, $V_i/2 > \overline{V}$, electrical power input decreases more steeply with temperature; starting with some axial temperature for which $P_e > P_c$, at some lower temperature P_e becomes less than P_r, the sign of $\nabla^2 T$ reverses; this results in constricted, bell-shaped, non-wall-stabilized temperature profiles.

The abruptness with which this can take place is illustrated by the results of numerical integration of such equations for a discharge in 2500 torr of mercury vapor with 25 torr of a gas having an average excitation potential of 4 eV, as illustrated in Fig. 18.

Calculations are repeated for various assumed values of ionization potential of the added gas, and ionization and excitation of mercury are neglected. For assumed ionization potentials between 6.5 and 7.0 volts, the temperature profile switches between wall stabilized and constricted. A number of such model calculations indicate the critical boundary is $\overline{V} = 0.585\ V_i$. This is different from the 0.5 previously mentioned because of temperature dependent pre-exponential factors in the P_e and P_r equations.

It turns out that many metals, especially those with rare-earth type spectra, have $\overline{V} < 0.585\ V_i$, while most common gases do not. Until the advent of metalhalide discharges, thermal equilibrium discharges in these metal vapors were not observed, because their vapor pressures are too low.

Metal halides as a class have vapor pressures of several tens of torr or more at temperatures accessible to quartz envelopes. Operation of high pressure thermal equilibrium arc in mercury vapor (or other relatively high ionization potential gas) in a quartz tube containing such halides results in vaporization of halide molecules from the walls and dissociation in the arc core. Thus the arc operates in a mixture of mercury, metal atoms and halogen atoms, and its temperature profile is substantially dominated by the ionization and radiation properties of the metal.

Since such discharge lamps can be made with nearly any metal in the periodic table, a considerable range of $\overline{V}/V_i$ is accessible. As already mentioned, metals with multiline, rare earth type spectra have $\overline{V}/V_i < 0.585$ and show constricted arcs in such discharges; other metals, such as the alkalis and two-electron metals, have $\overline{V}/V_i > 0.585$ and show wall stabilized arcs at the same total halogen content.

Lowke has reached similar conclusions to those outlined here from calculations for arcs containing dysprosium (21).

Similar calculations including the effect of absorption demonstrate that absorption of radiation can partially counteract a radiative constriction by transferring energy from the core to outer fringes of the arc. An optical depth of about three to ten appears optimum. For smaller optical thickness, too little radiation is absorbed in

FIG. 19. Composite photograph for arcs in 20 mm ID quartz tubing. Top, pure mercury; middle, mercury plus ThI_4; bottom, mercury plus ThI_4 plus NaI.

the outer fringes of the arc to affect its temperature; for greater optical thicknesses, too little radiation reaches the outer fringes of the arc to affect its temperature (22).

Empirically, it is found that radiatively-constricted arcs can be converted to wall-stabilized, by the addition of alkali metal halides to the arc tube contents. Incorporation of alkali metal reduces the effective ionization potential, while absorption in the strong resonance lines also helps eliminate constriction.

Fig. 19 is a composite photograph of three different arcs in the same diameter tube. First, a pure-mercury arc, wall stabilized; second, an arc in mercury plus thorium iodide; and third an arc in mercury plus thorium and sodium iodides.

V. SUMMARY

To summarize briefly, nearly every phenomenon known to the science of gaseous electronics occurs in discharge lamps of one form or another, beneficially or adversely. I have endeavored to outline a few of these for you. The reader who wishes more detailed information on these and other discharge lamp topics will find reference 23 of interest.

REFERENCES

1. M.A. Easley, J. Appl. Phys., 22, 590, (1951).
2. W. Verweij, Physica, 25, 980, (1959)
3. M. Koedam, A.A. Kruithof, and J. Riemens, Physica, 29, 565, (1963).
4. C. Kenty, J. Appl. Phys., 21, 1309, (1950).
5. W.E. Forsythe and E.Q. Adams "Fluorescent and Other Gaseous Discharge Lamps", Murray Hill Book Co., New York (1948), Table 11, p. 86.
6. L.J. Kieffer "Compilation of Low Energy Electron Collision Cross-Section Data, Part II, Line and Level Excitation", JILA Information Center Report No. 7, (Sept. 22, 1969).
7. B. Yavorsky, J. Phys. USSR, 10, 476, (1946).
8. J.F. Waymouth and F. Bitter, J. Appl. Phys., 27, 122, (1956).
9. M.A. Cayless, Proc. V International Conference on Ionization Phenomena in Gases, Munich, (1962).
10. J.F. Waymouth, Sylvania Technologist, 13, 2, (1960).
11. C.G. Found, Phys. Rev., 45, 519, (1934).
12. M.J. Druyvestyn and N. Warmoltz, Physica, 4, 41, (1937).

13. A.D. Forster-Brown and M.A. Cayless, Brit. J. Appl. Phys., 10, 409, (1959).
14. D.M. Speros and P.R. Bucilli, J. Electrochem Soc. 109, 940, (1962); ibid, 110, 748, (1963).
15. J.F. Waymouth, J. Appl. Phys., 30, 1404, (1959).
16. G. Medicus, J. Appl. Phys., 27, 1242, (1956).
17. J.F. Waymouth, J. Appl. Phys., 30, 1404, (1959) (Appendix).
18. J.F. Waymouth, Paper No. 1, M.I.T. Physical Electronics Conference, (1962).
19. D.A. Larson, H.D. Fraser, W.V. Cushing, and M.C. Unglert, Illum. Eng., 58, 434, (1963).
E.C. Martt, L.J. Smialek, A.C. Green, Illum. Eng., 59, 34, (1964) J.F. Waymouth, W.C. Gungle, J.M. Harris, and F. Koury, Illum. Eng. 60, 85, (1965).
20. J.F. Waymouth, Paper C5, High Pressure Arc Symposium, 22nd Annual Gaseous Electronics Conference, Oct. 1969.
21. J.J. Lowke, J. Appl. Phys., 41, 2588, (1970).
22. J.F. Waymouth, Paper C6, High Pressure Arc Symposium, 22nd Annual Gaseous Electronics Conference, Oct. 1969.
23. J.F. Waymouth "Electric Discharge Lamps", The M.I.T. Press, Cambridge, Mass. (1971) 338 pp.

Gaseous Electronics, eds. J.Wm. McGowan and P.K. John

GASEOUS ELECTRONICS IN THE UPPER ATMOSPHERE - SOME RECENT OBSERVATIONS OF THE ATOMIC OXYGEN 5577 and 6300 Å EMISSIONS

G.G. SHEPHERD

Centre for Research
in Experimental Space Science
York University
Toronto, Canada

I. INTRODUCTION

Gaseous electronics has been an integral part of upper atmospheric studies ever since Marconi successfully transmitted radio signals across the Atlantic. However, it was only after the pioneering analyses of Sidney Chapman on photodissociation of molecular oxygen (1) and the ionization processes arising from solar radiation (2) that the crucial role of photochemistry in the upper atmosphere became known. It is difficult to prescribe a comprehensive set of reviews in what has become a very broad field, but several recent books (3-5) give a good indication of the present status of the field. Good semi-annual reviews are also provided by the NATO Advanced Study Institute (6,7).

The gaseous electronics of the upper atmosphere is largely concerned with energy degradation processes. Energy inputs are largely provided by solar x-ray and UV radiation, and by auroral particle deposition. More subtle sources are provided by magnetosphere-ionosphere coupling, and by neutral motions of the upper atmosphere. In general, the microscopic processes are now reasonably well measured and understood, and it is the medium or larger scale processes that are now receiving the most attention. In this work, the study of optical emissions is a very direct way of monitoring energy degradation processes.

For this reason, observing auroral and airglow emissions has been a valuable way of studying the upper atmosphere, and still is an active field of endeavour. Contrary to what one might expect, the emissions which have been known the longest are still receiving most of the attention. This is partly because such observations acquire a diagnostic value only after they are understood, and partly because the excitation mechanisms for the well known emissions are as yet only partially understood.

In this paper, the excitation mechanisms for the forbidden atomic oxygen transitions at 5577 Å and 6300 Å are described and illustrated with recent upper atmospheric data. These emissions are well described by the opening remarks, and the description of them may be taken as illustrative of much of the state of upper atmospheric gaseous electronics today.

II. EXCITATION OF THE OI 5577 Å EMISSION IN THE AURORA

This is the most prominent emission in the aurora. Its identification began as a mystery (8) and its excitation remains a mystery today (9), although periods occurred when it was considered understood.

The nature of the mystery (10) is illustrated in Fig. 1, which shows photoelectric recordings of auroral intensity that contain characteristic pulsations. These auroral intensity pulsations, having quasi-periods of 5-10 seconds are a common and well known phenomenon whose origin is a mystery of another type. Two simultaneous records are shown in the figure, one arising from the 5577 Å emission, the other from the allowed N_2^+ 1st Negative Band at 3914 Å. The slight differences between the two traces can readily be explained by the differences in radiative lifetimes, about 0.7 sec and 10^{-8} sec respectively. The mystery arises from the similarity between the traces, and requires a detailed explanation. From the figure we might conclude that the excitation rates of the two emissions are closely linearly related, and many other observations bear this out. The N_2^+ emission is known to be produced by electron impact on ground state N_2, and its intensity is considered to be accurately proportional to the ionization rate (11). But the excitation of OI to the ^1S level involves no fewer than four processes:

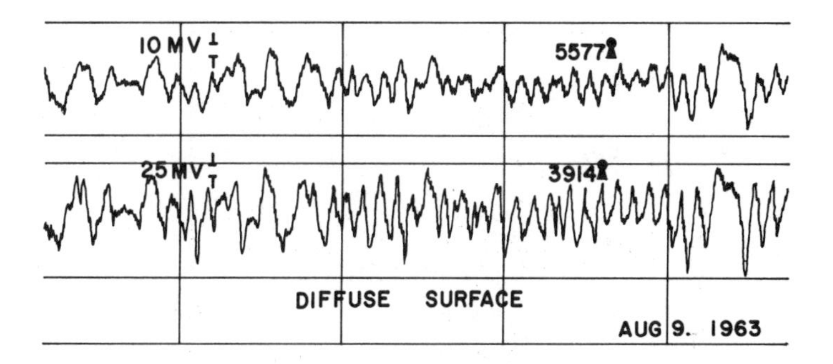

FIG. 1. Simultaneous recordings of auroral pulsations measured for the OI 5577 Å emission and the 3914 Å N_2^+ emission, recorded at Saskatoon by Dr. K.V. Paulson (1963). The time interval between major vertical rulings is 30 seconds.

$$e(\text{energetic}) + O \rightarrow O(^1S) + e \;.......1$$

$$e(\text{thermal}) + O_2^+ \rightarrow O(^1S) + O \;.......2$$

$$e(\text{energetic}) + O_2 \rightarrow O(^1S) + O + e \;...3$$

$$N_2(A) + O \rightarrow N_2(X) + O(^1S)....4$$

The relevant levels of N_2 and O are shown in Fig. 2. Process 1 was for many years considered to be the source of the 5577 Å line in the aurora, and only recently was the measured cross section, taken together with in-situ measurements of secondary electron flux shown to be inadequate to explain the emission (12). Process 2 is much better understood, and is certain to play a role in the aurora; it seems to be responsible for a fraction of the total emission, particularly that at high altitude. Processes 3 and 4 have been proposed more recently as solutions (13, 14). Brekke (15) and Henriksen (16) have evidence for 4, but one must conclude that at the present time neither has been confirmed. So far we have considered only the production mechanisms, and quenching must be introduced to determine the actual 5577 Å emission.

$$O(^1S) \rightarrow O(^1D) + h\nu(5577) \;...........5$$

$$O(^1S) + N_2 \rightarrow O(^3P) + N_2^* \;...........6$$

$$O(^1S) + NO \rightarrow O(^3P) + NO^* \;...........7$$

$$O(^1S) + O_2 \rightarrow O(^3P) + O_2^* \;...........8$$

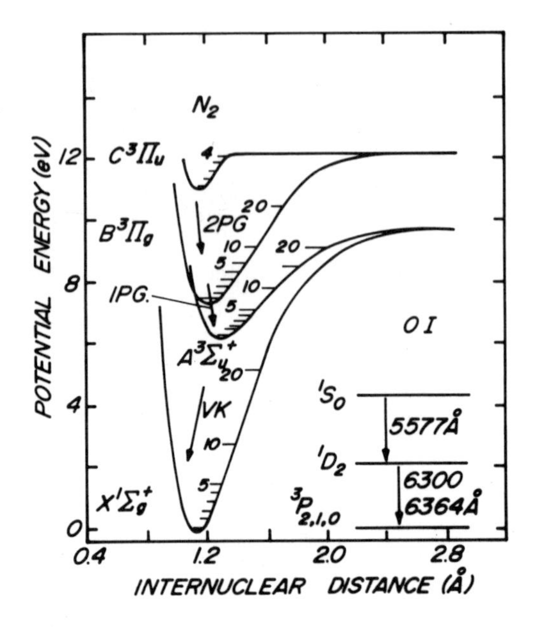

FIG. 2. Illustrating the energy levels of the N_2 molecule and the O-atom referred to in this paper.

The quenching by NO has recently become important as a result of measurements of Zipf, Borst and Donahue (17) of high NO concentration in the aurora. The mystery is then that these eight processes must be put together in such a way that the 5577 Å emission is closely proportional to the incident electron flux. It doesn't seem likely, and yet that is what was shown in Fig. 1.

III. OBSERVATIONS OF THE OI 5577 Å EMISSION IN THE AURORA

The data to be presented in this section were obtained (18) with a Wide Angle Michelson Interferometer, shown schematically in Fig. 3. This device is a Michelson interferometer that has an excess of refracting material in one beam. The result is that the ray traversing that beam has a virtual reflection point which can be adjusted (by proper location of the front surface mirror in the other beam) to coincide with the virtual image of that front surface mirror. Under these conditions the ray paths in both beams are

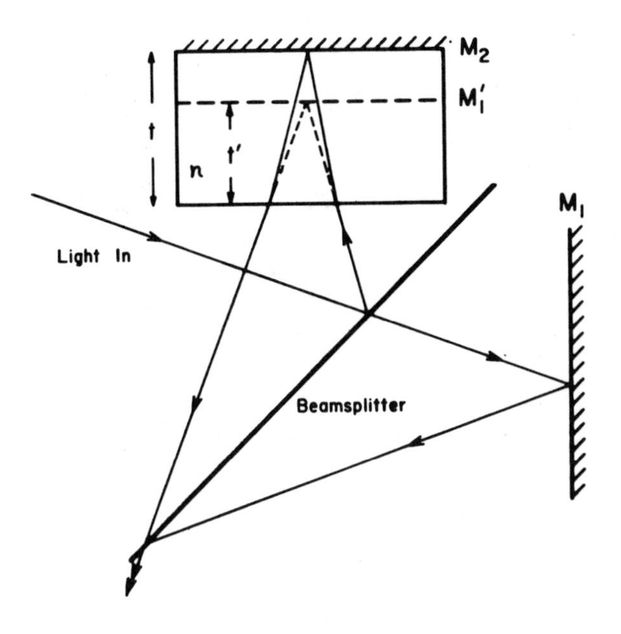

FIG. 3. A schematic drawing of the wide-angle Michelson interferometer. A slab of thickness t, inserted in one beam produces a virtual reflection at depth t', yielding symmetrical ray paths and a quasi-zero path difference condition.

identical, and the path difference is zero, independently of the incident angle. But the path difference is only quasi-zero; it is not zero because the travel times for the two beams are different. What this means is that one can adjust the device for a path difference of 10^6 fringes, yet accept a solid angle that is enormous - a result enjoyed by the ordinary Michelson interferometer only at zero path difference. To apply this instrument to the 5577 Å line, one oscillates the front surface mirror a few fringes and measures the fringe modulation depth - what Michelson called the fringe visibility. If the 5577 Å line is assumed to have a gaussian shape then this measurement alone is sufficient to yield the Doppler temperature of the emission. Using this technique it has been found (19) that during pulsations the temperature changes as well as the intensity. Since the temperature varies roughly inversely with the intensity it cannot be the result of heating. Zwick and Shepherd(20) confirmed this effect and some of their data is shown in Fig. 4, in which the time histories for several pulsations are

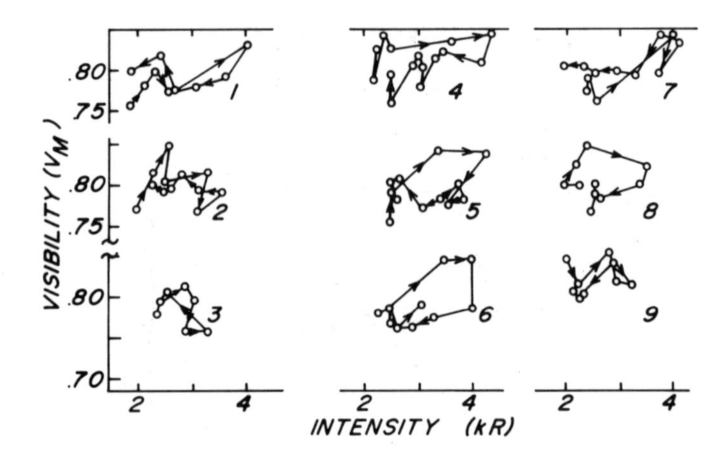

FIG. 4. Time histories of auroral pulsations. The sampling interval for a given pulsation, corresponding to the open circles, is 0.6 sec. Fringe visibility, which is inversely related to temperature, is plotted versus intensity. Pulsations 5, 6 and 8 are good examples of "clockwise" pulsations.

shown as plots of fringe visibility (which increases with decreasing temperature) against auroral brightness. Several identifiable patterns result, but the most frequently observed pattern is a clockwise rotation. For it, the visibility rises (temperature falls) with only a little change in intensity, and then the intensity rises subsequently, followed by a rise in temperature. The interpretation is as follows. The temperature rise is caused by the arrival of a burst of auroral electrons, having an energy considerably higher than that producing the auroral background level. These penetrate more deeply into the atmosphere and give rise to a lower observed temperature. The less energetic electrons which arrive a little later cause the subsequent increase in intensity. These data can be converted to plots of electron flux vs. time and electron energy vs. time, and one of these is shown in Fig. 5. The behaviour is consistent with an electron source at the equator, on a field line passing through the auroral zone which generates a simultaneous population of energetic electrons that disperse on their way to the earth. These conclusions have little to do with the knowledge of the excitation process for the 5577 Å line, and illustrate that optical measurements may lead to information on large scale processes as well as on microscopic ones.

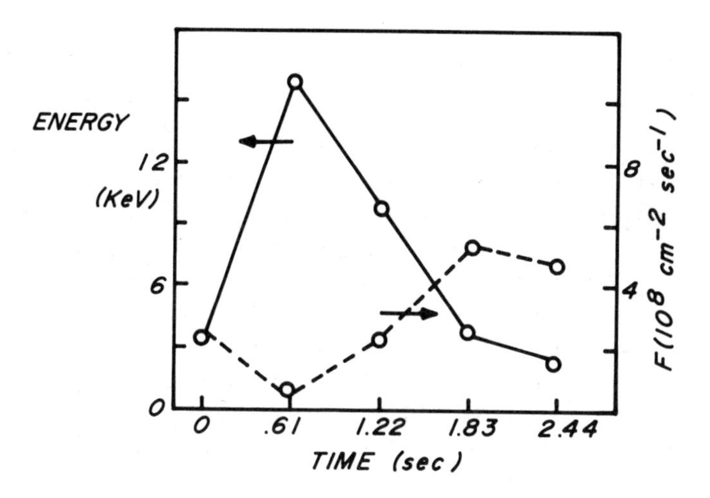

FIG. 5. The deduced temporal variation of auroral electron flux and auroral electron energy, for a clockwise pulsation.

IV. EXCITATION OF THE OI 6300 Å EMISSION IN THE AURORA AND AIRGLOW

The upper state $O(^1D)$ for the 6300 Å line is the lower state for the 5577 Å line. Hence every 5577 Å photon corresponds with the production of one ^1D atom, and all processes that lead to the 5577 Å line also lead to the 6300 Å line. This can also happen directly, as process 2, for example can leave an atom directly in the ^1D state. An important mechanism that is unique to the 6300 Å line is thermal excitation by hot electrons. When the electron temperature exceeds about 3500° K there are sufficient 2 eV electrons in the tail of the energy distribution to produce 6300 Å emission. In fact, with the available production processes, the 6300 Å emission should be incredibly bright; the fact that it is not is because of very severe quenching by molecular N_2, so severe that emission from below 200 km is practically negligible. This altitude restriction, along with the long radiative lifetime of 110 secs means that temporal variations of the 6300 Å emission bear little relation to those of the N_2 emission in aurora. Since this is what we expect, there is no mystery, and the assumption is that 6300 Å excitation is understood.

V. THE GLOBAL DISTRIBUTION OF 6300 Å EXCITATION

The various excitation processes of the 6300 Å emission are well illustrated by its global distribution, which is presently being mapped by York University Red Line Photometer on board the ISIS-II spacecraft. This joint U.S.-Canadian spacecraft was launched April 1, 1971 as the fourth of the series of International Satellites for Ionospheric Studies. There are two optical experiments on board, the Auroral Scanning Photometer (21) and the Red Line Photometer (22). Data primarily from the RLP will be shown here, though Dr. Anger's publications should be looked at to see his spectacular auroral pictures (23,24). The RLP has two optical inputs, 180° apart, both directed perpendicular to the spin axis of the spacecraft, which rotates once every 18 sec, approximately, in its near-polar near-circular orbit at 1400 km. The optical inputs are characterized by a 10 Å half-width filter, and an 88 Å half-width filter, such that they have equal response to white light, but a response ratio of about 9:1 for 6300 Å emission. Both inputs are combined onto one photomultiplier so it is the intensity sum that is telemetered, but since one channel scans the dark sky while the other scans the earth, the 6300 Å emission rate and the white light background can be substracted.

Fig. 6 shows how the data are presented. For each rotation of the spacecraft, and one (double) scan across the earth one line of print is generated by the computer. The 40 columns of three-digit numbers correspond to equal spacecraft rotation angles, giving emission intensities in units of 10R, and the leading edge of the earth as seen by the spacecraft is aligned with the left edge of the page. The trailing edge is arranged to fall somewhere within the right hand edge of the page. Successive lines of print are generated for successive spacecraft rotations as the spacecraft moves along in its orbit, so that its raster-like scan builds up a global picture of the 6300 Å emission. For mapping purposes the spin axis is held approximately in the orbital plane, so that the scanned strips are nearly perpendicular to the orbit plane but since the spin axis is fixed in inertial space, distortion occurs except for the point where the spin axis is parallel to the earth's surface; it increases progressively until the scanning plane walks off the edge of the earth. This is manifested by the curved right hand limb in the map. Rather than transforming these data to some other type of coordinates, the coordinates of each spatial resolution element in the spin map are calculated as separate computer print-outs that can be directly overlaid. There are many coordinates that one wishes to overlay in this fashion. In Fig. 7 the same data are shown again, but as contours of isointensity on which tracings of the overlays for magnetic dipole latitude and magnetic local time have been placed. From this it is evident that

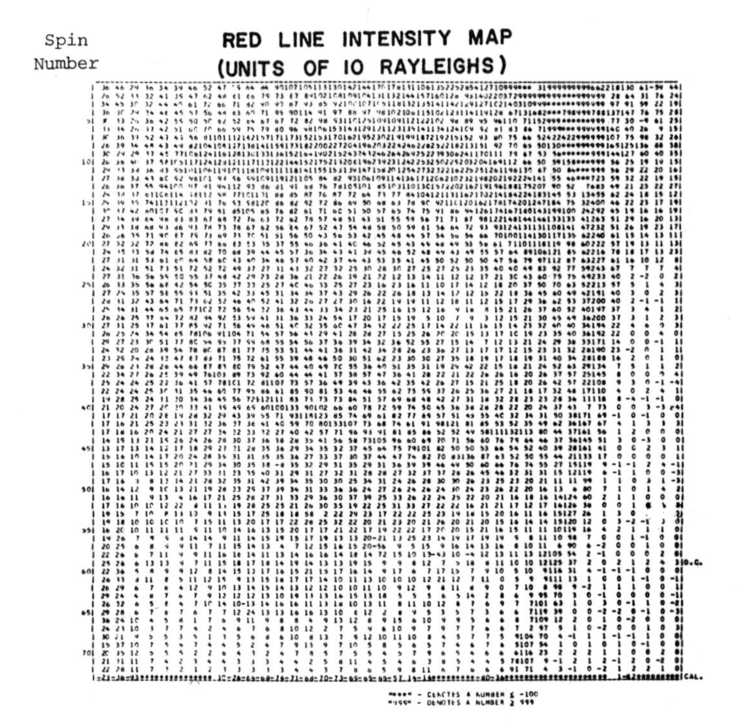

FIG. 6. A number map of the global 6300 Å emission, obtained from the red line photometer experiment on ISIS-II. One row of numbers corresponds to one spacecraft rotation, so that the picture is built up from 73 spins. The dayside aurora is visible near spin 10 and the nightside aurora near spin 40.

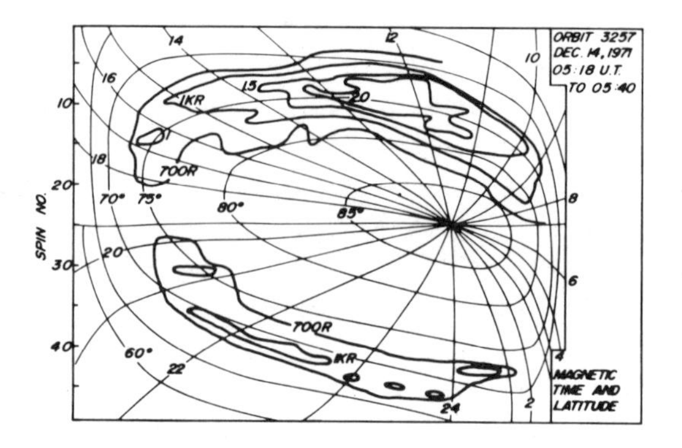

FIG. 7. Isointensity contours of the 6300 Å emission, traced from the data of Fig. 6 with overlaid coordinates of magnetic dipole latitude, and magnetic local time.

the emission is particularly bright at local noon, on the dayside of the earth. The data for this orbit have been described by Shepherd and Thirkettle (25).

This emission pattern is dramatically different from that for 5577 Å, for which a spin map derived from Dr.C.D. Anger's data for an orbit 2 days later is shown in Fig. 8. In this Figure (a) is the 6300 Å data for this Dec. 16, 1971 orbit; a "shading map" representation while (b) is the same for the 5577 Å emission. It is clear that the nightside 5577 Å auroral intensity is similar to that for the 6300 Å emission, but the 5577 Å dayside emission is practically absent. The difference between nightside and dayside is presumably one of auroral electron energy. Fig. 9 shows a contemporary model of the magnetosphere, showing how different the trajectories must be for dayside and nightside particles. Nightside electrons reach the earth via the magnetospheric tail, which can be entered only by some slow diffusion process. Once inside the tail, in a region known as the plasma sheet, the electrons are found energized from which they reach the nightside auroral oval with energies of about 5 keV. Acceleration may occur between the plasma sheet and the earth, but this matter is unclear as are many aspects of the nightside aurora. This is despite the fact that until recently it is the nightside aurora that has received practically the sole attention of auroral physicists.

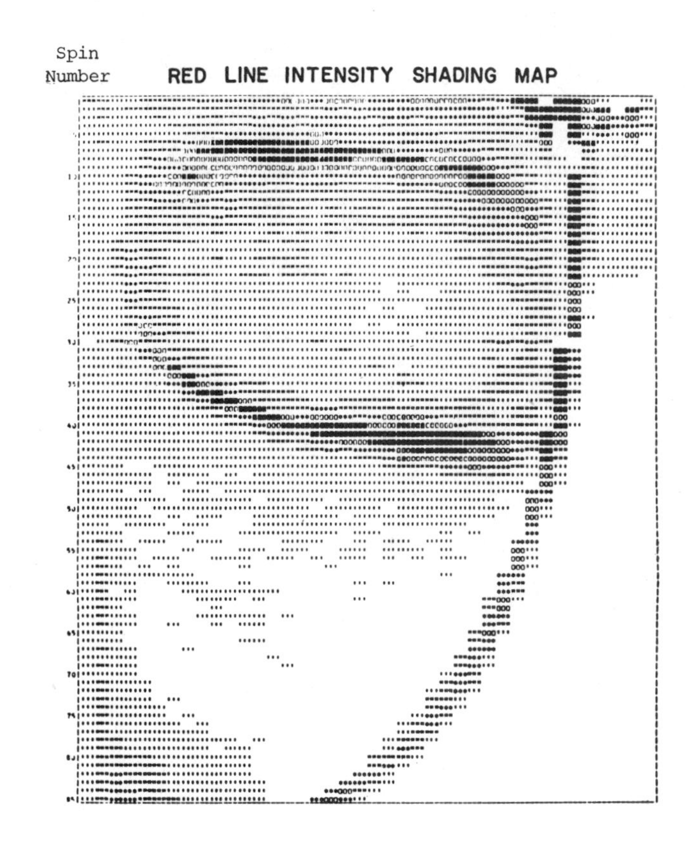

FIG. 8. (a) The 6300 Å data for Dec. 16, 1971, in precisely the same format as for Fig. 6, but for which intensity is represented by the printed symbol density.

The dayside aurora has only recently become, known and yet its origins may be simpler. The dayside oval can be considered to connect directly to the magnetosheath, a region of shocked solar plasma, lying just inside the bow shock, defining the outer edge of the earth's magnetosphere. There are a number of theoretical problems associated with "direct" entry through the dayside magnetospheric cusp, and only the experimental evidence has made the process credible. But from a simple-minded point of view it seems reasonable that the dayside auroral electrons of 100-200 eV energy enter directly without acceleration from the magnetosheath. The very large I(6300)/I(5577) ratio on the dayside is consistent with this.

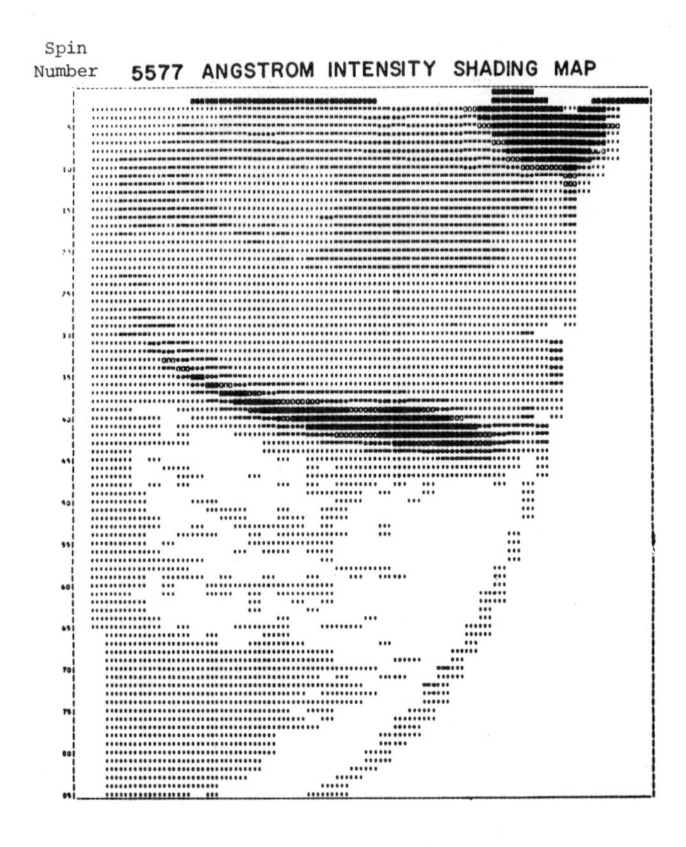

FIG. 8 (b) The same as (a), but for the 5577 Å emission, from data provided by Dr. C.D. Anger. Note the difference between day-side (spin 7) and nightside (spin 41) aurora for the two emissions.

Now to return to the theme of 6300 Å excitation. On the dayside the excitation probably occurs through direct electron impact, excitation by secondary electrons, excitation by ambient hot electrons, and some by dissociative recombination. All of this must occur at sufficiently high altitude (about 400 km) that quenching of ^1D atoms does not make the 5577 Å emission competitive in intensity. In the nightside aurora the excitation appears to be primarily from the secondary electrons produced by the 5 keV electrons at 100-200 km. At these levels a large fraction of the 6300 Å emission is quenched and the 5577 Å emission is competitive or even dominant. Some emission would arise from dissociative recombination in the enhanced auroral ionosphere.

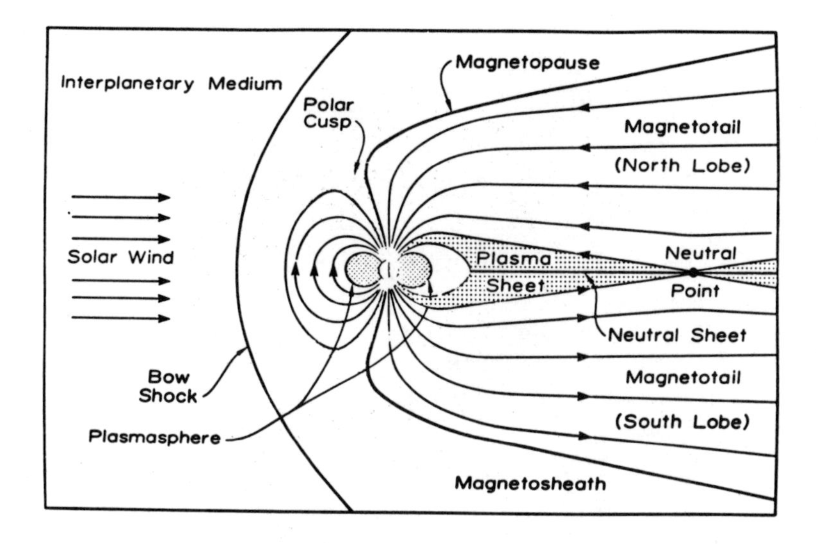

FIG. 9. A current model of the magnetosphere, showing the entry point for dayside aurora (dayside cusp) and the plasma sheet in the magnetotail that is associated with the nightside auroral source.

Returning to Fig. 6 one sees a fringe of weaker emissions lying equatorward of the auroral oval. Its termination is manifested by an intensity drop from 300 R to about 50 R. This 300 R region is characterized by the fact that the conjugate ends of the magnetic field lines are sunlit in the summer southern hemisphere. The interpretation is that the emission is produced by photoelectrons generated in the opposite hemisphere, which travel along the field lines, with little loss of their 20-50 eV energy. The excitation process is likely by direct photoelectron impact on atomic oxygen..

In Fig. 10 a spin map is shown for dramatically different circumstances, during a major geomagnetic storm. Evident features are the enormous expansion of the auroral emission over the polar cap, and the disappearance of the conjugate photoelectron intensity step, and a new feature just equatorward of the aurora, a mid-latitude red arc, often called a SAR (stable auroral red) arc. This phenomenon, which has been shown to be purely 6300 Å emission (apart from a few isolated reports of other emissions), is now known to arise by purely electron thermal excitation, through heat conduction in the electron gas from an equatorial energy source associated with the

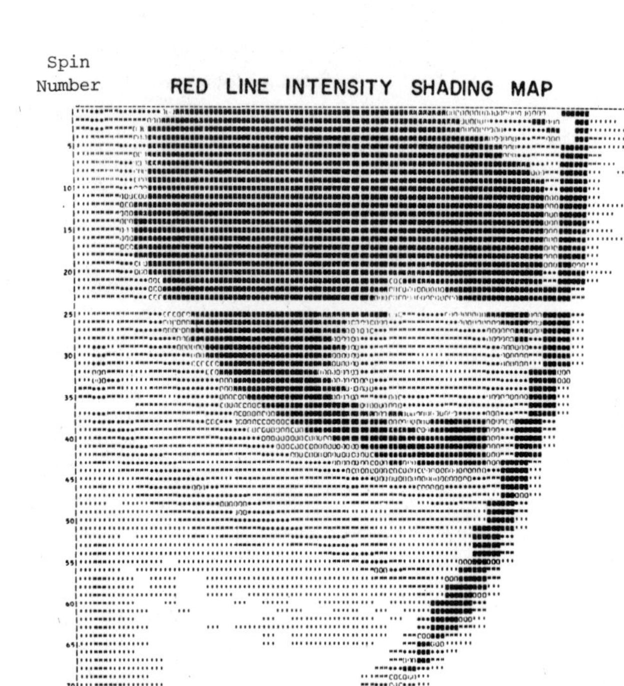

FIG. 10. The global 6300 Å emission pattern existing during a major geomagnetic storm on Dec. 18, 1971

storm-time ring current.

Near the bottom of the map an intensity enhancement is obvious. This is associated with the equatorial electrojet, and results from an enhanced rate of dissociative recombination. Thus many distinct excitation processes can be envisaged in this global pattern. This demonstrates that the viewing of large scale processes may help in the understanding of microscopic processes.

REFERENCES

1. S. Chapman, 1930. Phil. Mag. 10, 369.
2. S. Chapman, 1931. Proc. Phys. Soc. (London) 43, No. 1, 26.
3. H. Rishbeth, 1969. Introduction to Ionospheric Physics, Academic Press, New York.
4. J.A. Ratcliffe, 1972. An Introduction to the Ionosphere and Magnetosphere, Cambridge University Press.
5. R.C. Whitten and I.G. Poppoff, 1971. Fundamentals of Aeronomy, John Wiley and Sons, Inc., New York.
6. B.M. McCormac and A. Omholt, (ed.), 1969. Atmospheric Emissions, Van Nostrand Reinhold Company, New York.
7. B.M. McCormac, 1971. The Radiating Atmosphere, D. Reidel Publishing Company, Dordrecht-Holland.
8. J.C. McLennan and G.M. Shrum, 1925. Proc. Roy. Soc. A, 108, 501.
9. G.G. Shepherd, 1972. Ann. Geophys. 28, 99.
10. K.V. Paulson, 1963. Fluctuations in Brightness from Quiet-Form Aurorae, Ph.D. Thesis, University of Saskatchewan, Saskatoon.
11. M.H. Rees, 1963. Planet. Space Sci. 11, 1209.
12. T.M. Donahue, T. Parkinson, E.C. Zipf, J.P. Doering, W.G. Fastie and R.E. Miller, 1968. Planet. Space Sci. 16, 737.
13. T.D. Parkinson, E.C. Zipf and T.M. Donahue, 1970. Planet. Space Sci. 18, 187.
14. T.D. Parkinson and E.C. Zipf, 1970. Planet. Space Sci. 18, 895.
15. A. Brekke, 1973. Planet. Space Sci. 21, 698.
16. K. Henriksen, 1973. Planet. Space Sci. 21, 863.
17. E.C. Zipf, W.L. Borst and T.M. Donahue, 1970. J. Geophys. Res. 75, 6371.
18. R.L. Hilliard and G.G. Shepherd, 1966a. Planet. Space Sci. 56, 362.
19. R.L. Hilliard and G.G. Shepherd, 1966b. Planet. Space Sci. 14, 383.
20. H.H. Zwick and G.G. Shepherd, 1973. Planet. Space Sci. 21, 605.
21. C.D. Anger, T. Fancott, J. McNally and H.S. Kerr, 1973. Appl. Opt. 12, 1753.
22. G.G. Shepherd, T. Fancott, J. McNally and H.S. Kerr, 1973. Appl. Opt. 12, 1767.
23. C.D. Anger and A.T.Y. Lui, 1973. Planet. Space Sci. 21, 873.
24. A.T.Y. Lui and C.D. Anger, 1973. Planet. Space Sci. 21, 799.
25. G.G. Shepherd and F.W. Thirkettle, 1973. Science 180, 737.

Gaseous Electronics, eds. J.Wm. McGowan and P.K. John

6 VIBRATIONAL EXCITATION OF MOLECULES VIA SHAPE RESONANCES

G. J. SCHULZ
Department of Engineering and Applied Science
Mason Laboratory
Yale University, New Haven, Connecticut

I. INTRODUCTION

This paper describes very briefly the progress made in the past decade or so in our basic understanding of electron-impact cross sections at low energy, with special emphasis on vibrational excitation of molecules via resonances. The need for having reliable electron-impact cross sections for vibrational excitation has been recognized for a long time. Many phenomena in gaseous electronics cannot be understood without having these cross sections. Thus, researchers in gaseous electronics have looked to experimentalists and theorists who concerned themselves with electron cross sections for guidance. The response of this group, however, was not always enthusiastic. The electron-beam fraternity usually had different roots. The early electron beam experiments were at the heart of Physics and led directly to the development of atomic spectroscopy and to quantum mechanics and thus these groups were concerned with the fundamental laws of nature rather than with some strange phenomena in discharges. Theoreticians were especially vocal in concerning themselves with the fundamental problems of collision theory and they often limited their interest to an analysis of atomic hydrogen. Carbon dioxide and nitrogen, two important species in gaseous electronics, were far from their capabilities.

In recent years we have found that the molecules of interest to gaseous electronics (and also the upper atmosphere) have most interesting properties, which make their detailed study worthwhile and fascinating, for their own sake and for the sake of interesting applications.

The aspect which makes the cross sections for low-energy electrons on molecules both fascinating and understandable from a theoretical viewpoint is the existence of "resonances" or "compound states" at more or less well-defined energies (1). The existence of "resonances" has been understood in nuclear physics for many decades, but in molecular collisions the concept has only received attention in the past decade.

II. ENERGIES OF COMPOUND STATES

Figs. 1 and 2 show the energy levels of most of the presently known resonances in N_2 and CO, respectively. These energy level diagrams are fairly typical of many diatomic molecules and even of triatomic and more complicated systems. At low energies (0 - 4 eV above the ground state), one or more compound state exist.

The extra electron which forms these states is held to the molecule

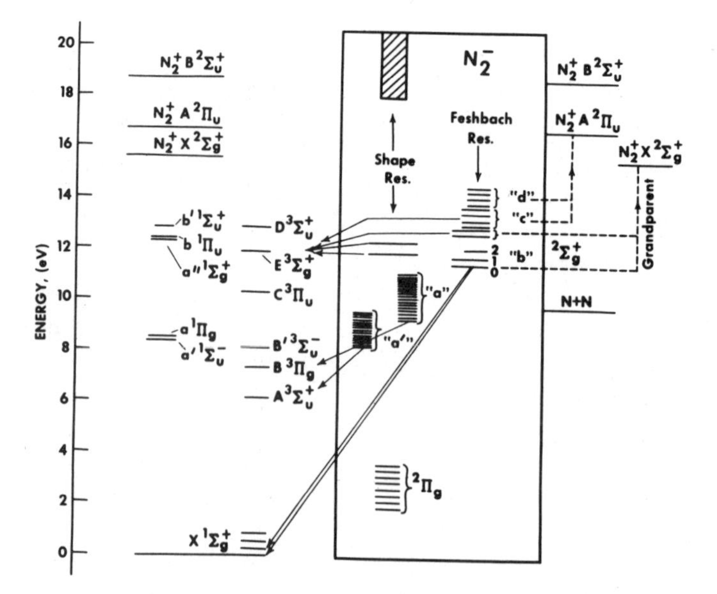

FIG. 1. Energy level diagram of N_2 and N_2^- in schematic form (1). The energy levels in the boxed portion are compound states and are divided into shape resonances and Feshback resonances. Some decay channels are indicated with arrows. The dashed lines leading to the positive ion states indicate the "grandparents" of the appropriate Feshback resonances.

by a barrier caused mostly by the centrifugal potential. We call these types of states "shape resonances". We shall discuss their importance below, in fact we shall concentrate on these shape resonances in this review, since they influence the distribution functions and energy loss processes in discharges to a significant extent.

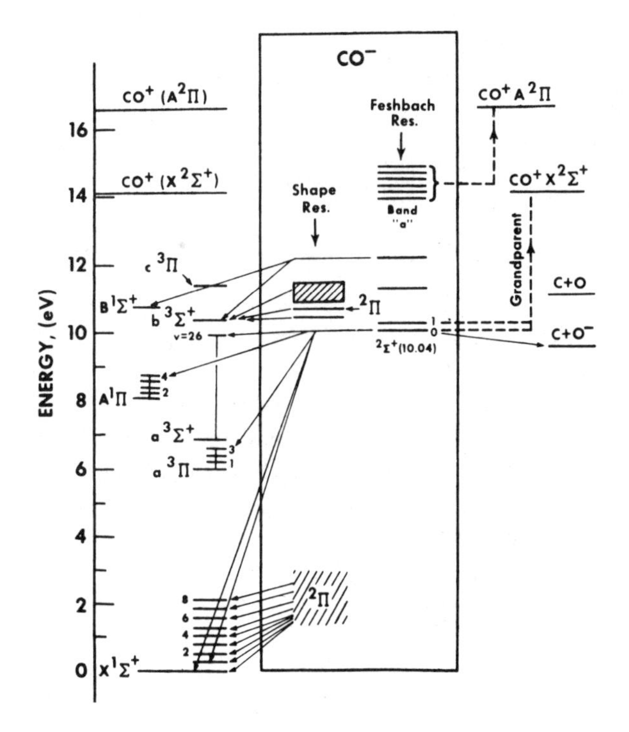

FIG. 2. Energy level diagram of CO and CO^- (1).

Shape resonances can be associated not only with the ground electronic state but also with valence excited states (see bands "a" and "a'" in Fig. 1) and with Rydberg excited states. No matter with which state a shape resonance is associated, it always lies above the state from which it derives. All these shape resonances lead to vibrational excitation of the nuclei.

At higher energies--starting below the first Rydberg excited state of the molecule--we find resonances which lie below the state from which they derive. (see e.g. band "b" in Fig. 1) We call these resonances Feshbach resonances and they are so designated in Fig. 1 and 2. These resonances often dominate the threshold behavior of the excitation cross sections to electronically excited states and cause structure in the elastic cross section.

Resonances associated with doubly excited states of atoms and molecules also exist. These states can lead to vibrational excitation as well as structure in the ionization cross section and the electronic excitation.

III. SHAPE RESONANCES AT LOW ENERGY (0 - 4 eV)

Diatomic molecules such as H_2, N_2, CO, NO, O_2 exhibit shape resonances at low energy which lead to large cross sections for vibrational excitation. We know that some triatomic molecules (2), specifically CO_2 and more complex molecules (e.g.; benzene) also have such shape resonances. The lifetime and the energy of these shape resonances determine the behavior of the vibrational cross section, and we discuss the role of the lifetime on the behavior of the vibrational cross section below. In this section we are concerned with the lowest shape resonance, i.e. the lowest state of the XY^- ion, consisting of the ground state of the molecule plus an electron in the lowest unfilled orbital. Table 1 shows the molecular orbitals involved in the case of N_2, for the first shape resonance, the valence excited shape resonance and the higher resonances.

A. Short Lifetime

For the shorter lived resonances in which the lifetime is comparable with the transit time of the electron across the molecular diameter, the various vibrational cross sections exhibit broad featureless humps, located around the energy at which the compound state exists. A further feature of this category of compound states is the rapidly diminishing ratio of the magnitudes of the vibrational excitation cross sections to successive vibrational levels. Such resonances represent the short-lifetime limit of compound states, generally described as the impulse limit beyond which the resonance and direct processes are indistinguishable. The 3 - 4 eV shape resonance in hydrogen offers an example of this class of resonance.

B. Intermediate Lifetime

Once the lifetime of the compound state significantly exceeds the time required by the electron to travel directly across the molecular diameter, a "standing wave" is set up in the compound state at more or less well-defined energies. This causes the vibrational cross sections to develop an oscillatory energy dependence. The location of the peaks and valleys comprising these oscillations shifts toward higher energies as a function of increasing vibrational quantum number of the final state. This behavior is demonstrated by the first shape resonance in N_2 and CO, located at 2.3 eV and 1.7 eV respectively, which possess lifetimes of the order 10^{-14} sec. (see Fig. 3)

The behavior of the vibrational cross section in the case of the 2.3 eV nitrogen resonance has been successfully explained by the Boomerang model proposed by Birtwistle and Herzenberg (3). The model proposes that only a single outgoing and a single reflected N_2^- nuclear wave exist, that is, the lifetime of the compound state is comparable with a vibrational period. The annihilation of the nuclear wave is caused by the emission of the extra electron. The energy of this outgoing electron depends on the instantaneous internuclear separation, R. Since the penetration through the potential barrier is a strong function of electron energy, the barrier penetrability becomes a function of R. Thus we see that the autoionization width is also a function of R. The magnitude of $\Gamma(R)$ is adjusted to provide the desired annihilation rate.

In the case of N_2, the compound state has a symmetry $^2\Pi_g$ and the decay of this N_2^- state leads to the excitation of 10 vibrational levels. The cross sections to v = 1-8 have been previously discussed (1) and here we show in Fig. 3 the cross sections to v = 7-10. Just as was the case for v = 1-8, the vibrational cross sections to v = 7-10 exhibit an oscillatory energy dependence, with the peaks shifting on the energy scale.

The absolute magnitude of the vibrational cross section at the first peak as a function of vibrational quantum number is shown in Fig. 4.

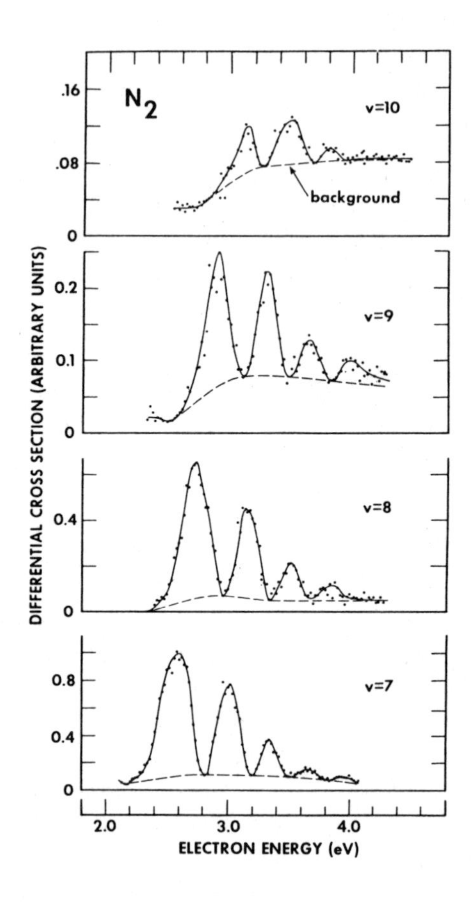

FIG. 3. Energy dependence of vibrational cross sections to v = 7-10 in N_2, on a compatible scale (4).

The higher vibrational states have small cross sections. Recently, Boness and Schulz (4) have shown that, for vibrational states larger than about 4, the dominant reason for the decrease of the vibrational cross section lies in the penetration of the potential barrier. When the compound state decays to <u>higher</u> vibrational states, it emits an electron of <u>lower</u> energy, lower energy electrons have great difficulty penetrating the barrier and hence the cross section to higher states is smaller.

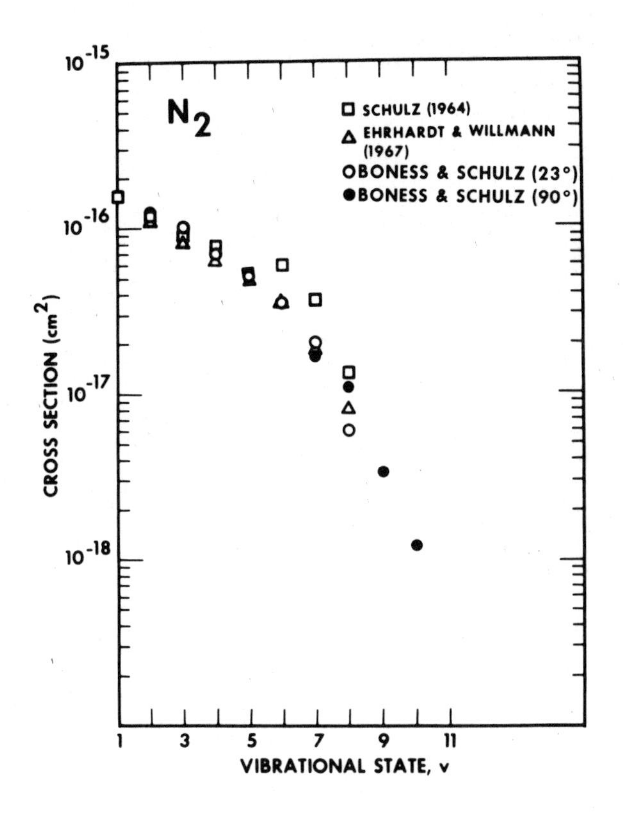

FIG. 4. Magnitude of the first peak in the vibrational excitation of N_2 vs. quantum number of vibrational state, v = 1-10 (4). All cross sections are referred to the value of Schulz (11).

C. Long Lifetime

With further increase in the resonance lifetime, one approaches the so-called compound molecular limit. The vibrational cross sections of those molecules possessing resonance lifetimes appropriate to this category exhibit very narrow isolated spikes. The positions of these spikes remain constant in energy for all final vibrational states and these locations correspond to the vibrational levels of the resonance. An example is oxygen in the energy range 0-1.6 eV. The energy-integrated cross sections in the energy range from 0.33 eV (corresponding to the v' = 18 of the compound state) is shown in Fig. 5. The cross section consists of "spikes" of narrow width (certainly less than 50 meV) at the position of the vibrational levels of the $^2\Pi_g$ compound state.

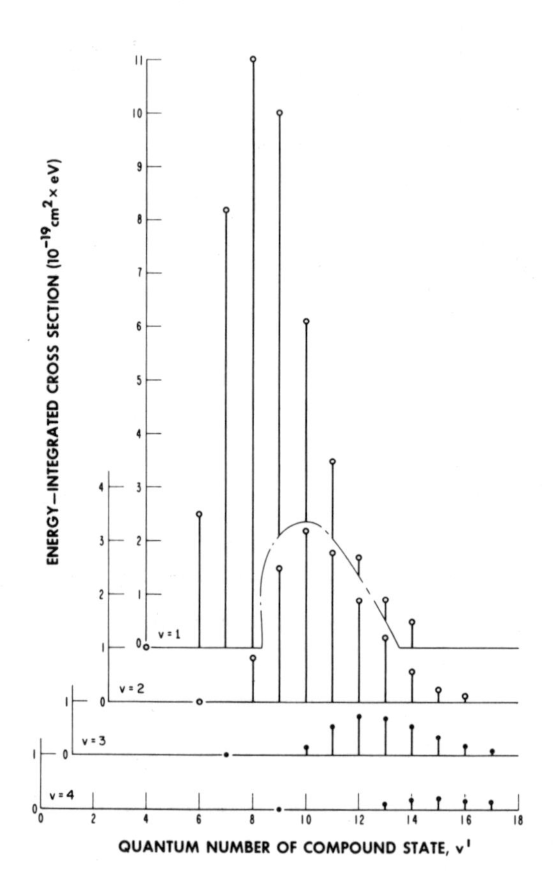

FIG. 5. Energy-integrated cross section for vibrational excitation in O_2 vs vibrational quantum number of compound state (12). The horizontal axis spans the energy from 0.330 eV (v' = 6) to 1.649 eV (v' = 18)

IV. SHAPE RESONANCES AT INTERMEDIATE ENERGIES

The resonances described in the previous section, namely the lowest states of the XY^- system, are not the only ones which exist. Molecules may have many additional states of the XY^- system, which are connected with the $X + Y^-$ or $X^- + Y$ state. Only a few of the many states, however, need to be considered in the present context. The XY^- states which can serve as intermediates for vibrational excitation must be accessible by electron impact on the ground state of the molecule and they must have an appropriate lifetime resulting from a potential barrier which confines the electron in the neighborhood of the molecule. These conditions are so stringent that only a single molecule studied to date has exhibited this effect, namely, oxygen.

Fig. 6 shows the cross sections obtained by Wong et al (5) for the excitation of the first few vibrational levels of the ground electronic state of O_2 in the energy range from 4 eV to 14 eV. The broad structureless curve indicates that the resonance responsible for this excitation mechanism is probably short-lived. Wong et al postulate that the appropriate resonance for this regime is the $^4\Sigma_u^-$ state of O_2^-.

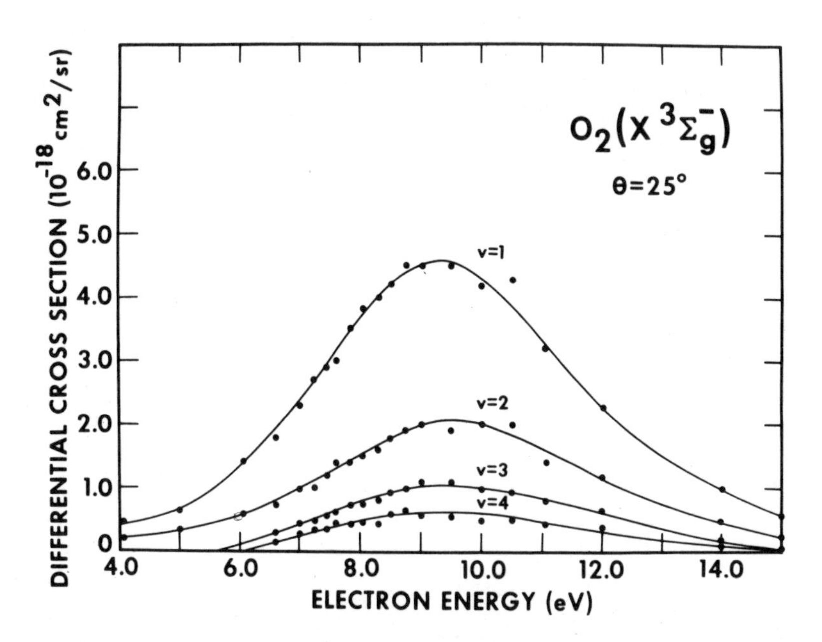

FIG. 6. Vibrational excitation of O_2 in the energy range 4-14 eV(5).

V. SHAPE RESONANCES ASSOCIATED WITH EXCITED STATES

Electronically excited states of molecules also can have associated shape resonances. Bands "a" and "a'" marked in Fig. 1 in the range 8-11 eV are examples of such shape resonances. These particular examples, for the case of N_2 are associated with the $A^3\Sigma_u^+$ and $B^3\Pi_g$ valence states of N_2. These shape resonances decay by the emission of an electron into vibrational levels (v = 1-6) of the electronic state from which they derive ($A^3\Sigma_u^+$ and $B^3\Pi_g$ respectively). This mechanism appears to be very efficient in populating vibrational levels of electronically excited species.

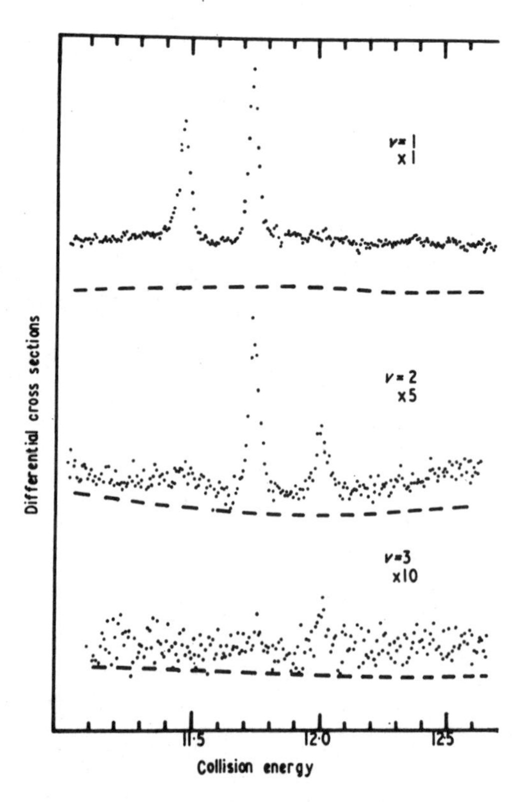

FIG. 7. Vibrational excitation via Feshbach resonances in N_2 (8).

Feshbach resonances such as band "b" in Fig. 1 can also populate vibrational levels of the ground electronic state. Fig. 7 shows the differential cross section for v = 1, 2 and 3 in N_2 as obtained by Comer and Read (8). The peaks of Fig. 7 are at the positions of band "b" shown in Fig. 1.

VI. RESONANCES ABOVE THE IONIZATIONAL POTENTIAL

Resonances above the ionization potential have been well documented for many atoms and molecules. There are many decay channels for such resonances, including vibrational excitation of the ground electronic state. The nitrogen molecule is an example: In the region 15 eV to 30 eV, a large vibrational cross section is evident for excitation of the ground electronic state as shown in Fig. 8. Pavlovic et al (9) proposed that the resonances involved consist of doubly excited valence states plus an extra electron (see Table 1).

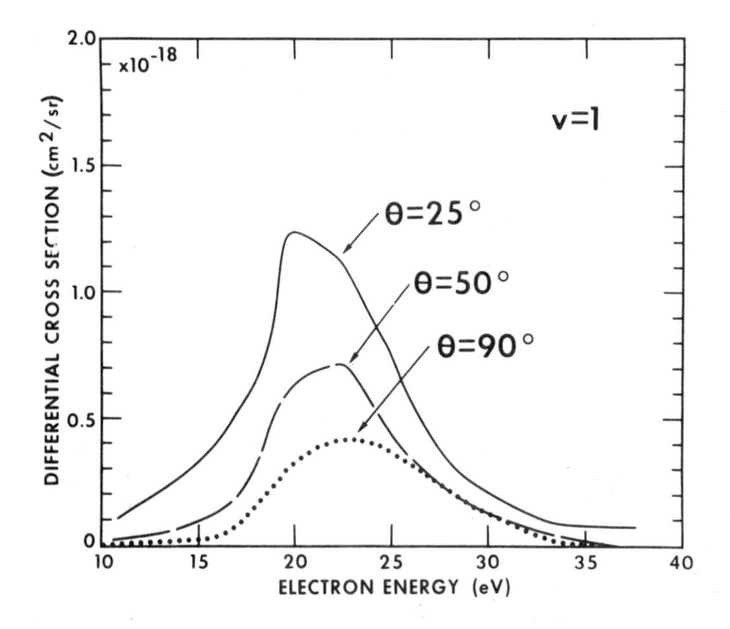

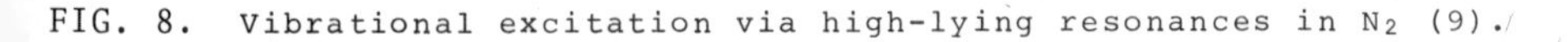
FIG. 8. Vibrational excitation via high-lying resonances in N_2 (9).

APPROXIMATE MOLECULAR ORBITALS FOR N_2 and N_2^-

		VALENCE ELECTRONS							RYDBERG ELECTRONS		STATE	ENERGY, (eV)
		KK	$(\sigma_g 2s)^2$	$(\sigma_u 2s)^2$	$(\sigma_g 2p)^2$	$(\pi_u 2p)^4$	$(\pi_g 2p)^4$	$(\sigma_u 2p)^2$	$(\sigma_g 3s)^2$	( )		
N_2	GROUND STATE	✓	✓	✓	✓	✓	0	0	0	0	$X^1\Sigma_g^+$	0
N_2^-	SHAPE RES.			ditto			1	0	0	0	$^2\Pi_g$	1.7
N_2	VALENCE EXCITED	✓	✓	✓	✓	-1	1	0	0	0	$A\,^3\Sigma_u^+$	6.2
N_2^-	CORE-EXCITED SHAPE RES.			ditto			1	1	0	0		8.2
N_2^+	POSITIVE ION	✓	✓	✓	-1	✓	0	0	0	0	$X\,^2\Sigma_g^+$	15.6
N_2	RYDBERG STATE			ditto			0	0	1	0	$E\,^3\Sigma_g^+$	11.87
N_2^-	FESHBACH RES.			ditto			0	0	2	0	$^2\Sigma_g^+$	11.48
N_2^-	CORE-EXCITED SHAPE RES.			ditto			0	0	1	1		11.92
N_2	DOUBLY EXCITED	✓	✓	✓	-1	-1	1	1	0	0		~20
N_2^-	DOUBLY EXCITED RES.			ditto			2	1	0	0		~20

TABLE 1

VII. SHAPE RESONANCES IN COMPLEX MOLECULES

With the exception of the CO_2 molecule, not much information is yet available in vibrational excitation in triatomic and more complex molecules. However, we know the location of low-lying shape resonances from various transmission experiments, which have recently been reviewed by Sanche and Schulz (2). In CO_2, NO_2 and C_6H_6, structure exists in the total scattering cross section, which we interpret as showing shape resonances of intermediate or long lifetime. Broad structures exist in the total scattering cross section in N_2O, H_2S and C_2H_4, indicating short-lived shape resonances. In CO_2, the shape resonance centers around 3.8 eV and its decay leads to the excitation of about 30 vibrational modes. Boness and Schulz (10) have classified the energy-loss spectrum (Fig. 9) into two progressions; namely, a progression of pure symmetric stretch modes, n00, which extends from 100 to 10, 0, 0 and a progression of symmetric stretch modes with one quantum of bending, n10, extending from 110 to 20, 1, 0. The energy dependence of the vibrational cross sections shows an oscillatory behavior similar to the case of N_2. Fig. 10 shows a sample of the vibrational cross sections vs electron energy.

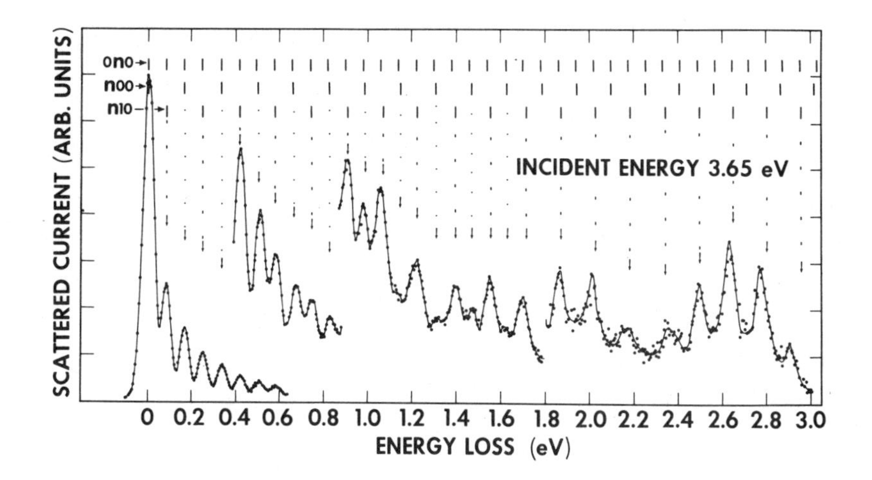

FIG. 9. Energy-loss-spectrum in CO_2 at an incident energy of 3.65 eV. The energy-loss peaks seem to include two series: n00 and n10 (10).

VIII. CONCLUSION

The author's involvement in the phenomena outlined above started 20 years ago with his thesis on microwave gas discharge, under the direction of S.C. Brown and W.P. Allis. It was then clear that, if further progress in our understanding of discharges should take place, one must have more reliable and more extensive cross sections for low-energy electrons. Information on vibrational excitation was then almost absent, but it was realized that these cross sections were badly needed. Although we do not yet know all that we would like to know about these effects, it is clear that great progress has been made in our understanding of a fascinating story and the newly gained knowledge is being fed back into the field of gaseous electronics, completing an interesting feedback loop in scientific information.

Throughout these exciting years, the Gaseous Electronics Conference, long under the enlightened direction of W.P. Allis, played a most central, sometimes catalytic role in motivating the appropriate interest and in promoting high quality in the field. Through the Gaseous Electronics Conference, the field has had a center for in-

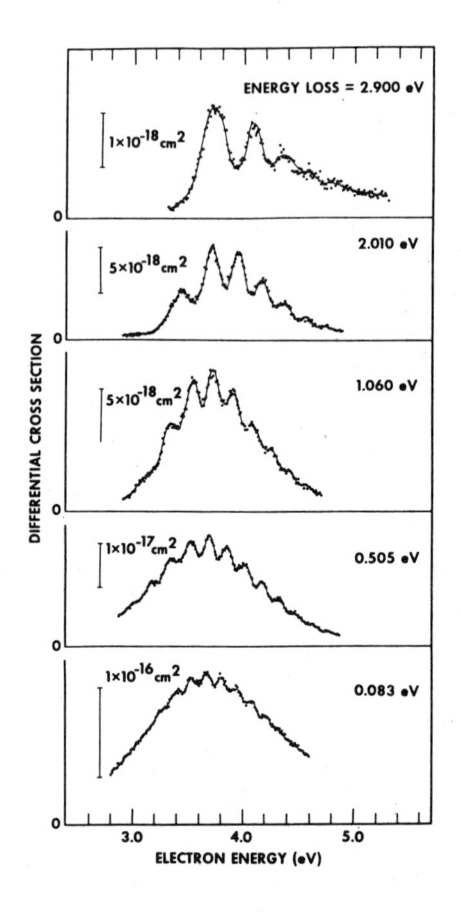

FIG. 10. Energy dependence of vibrational cross sections in CO_2 in the region of the first shape resonance (10).

formation exchange, arguments, agreements and a great deal of unity.

ACKNOWLEDGEMENT

This work was supported in part by the Army Research Office and the Defense Nuclear Agency under Subtask HD010 work unit 40, "laboratory reaction rate studies" and in part by the Office of Naval Research.

Gaseous Electronics, eds. J.Wm. McGowan and P.K. John

7 NON-EQUILIBRIUM EFFECTS IN ARC DISCHARGES

J. UHLENBUSCH

Physikalisches Institut II
der Universitat Dusseldorf
West Germany

I. INTRODUCTION

Plasmas generated in CW arc discharges (cascaded arcs) cover a broad range of temperature and electron density. Under certain conditions arc discharges are quite stable and have a well defined cylindrical geometry. Thus they are a good source for precision measurements of plasma properties over a wide range of parameters. As an additional advantage it is possible to check the measured data by solving the relevant plasma equations, where all state variables have a radial variation only.

This paper deals in the first part with the description of arc plasmas by thermodynamic methods, i.e. complete, local and partial equilibrium situations are discussed and compared with experiments. In the second part non-equilibrium effects due to high transport rates of particles are investigated. A method is given to calculate the state variables and some experimental data are reported. Finally in the third part criteria for the validity of a thermodynamic approach are discussed.

The last section deals with non-equilibrium effects caused by strong heat fluxes in the vicinity of the discharge walls. The modifications of the kinetic theory necessary to explain the phenomena are summarized and applied to arc discharges.

II. EQUILIBRIUM MODELS

The starting point for the theoretical description of an arc plasma is the kinetic equation. Suppose the plasma consists of particles of the kind s (e.g. hydrogen atoms, electrons, photons) in the quantum state q, where q is a set of all quantum numbers related to the excitation state in question. Let i be an index characterising the state of ionization or dissociation (e.g. H^+, He^{++}, etc.) and $\vec{V}_{sqi}$ the velocity of the particles at time t in a volume element d^3r situated at r. Then the number of particles of kind $\alpha = \{s,q,i\}$ in a small volume $d^3\vec{r}$ around $\vec{V}_\alpha$ and in a small volume $d^3\vec{V}_\alpha$ around $\vec{V}_\alpha$ (namely with a velocity between V_α and $\vec{V}_\alpha + d\vec{V}_\alpha$) at time t is given by

$$dN_\alpha = f_\alpha(\vec{r},\vec{v}_\alpha, t)d^3\vec{r}\ d^3\vec{V}_\alpha$$

where f_α, the density of particles with the discrete property α in a continuous phase space with the variables $\vec{r}$ and $\vec{V}_\alpha$ is the distribution function

The number of particles of species α in the volume element $d\tau = d^3\vec{r}\ d^3\vec{V}_\alpha$ can be changed in the time interval $d\tau$ by production and by loss of particles inside the volume element $d\tau$ and by particles flowing in or out through the surface of $d\tau$ thereby increasing or decreasing the number density f_α. Because there are negative (bound) and positive (free) energy states of atoms, molecules and electrons, the loss and gain processes inside the volume $d\tau$ can be divided into several groups.

1. Loss and production by elastic collisions between particles, including photons, namely collisions by which only the velocity (kinetic energy) is changed; all quantum numbers $\alpha = \{s,q,i\}$ remain constant.
2. Loss and production by inelastic collisions between particles including photons, namely collisions by which the kinetic energy as well as the quantum state q of bound states are altered.
3. Loss and production by reactive collisions between particles (including photons), namely collisions by which the kinetic energy as well as the quantum number q, the index i (dissociation, ionization) and the particle group s (nuclear reactions in a fusion reactor) are modified. Particles passing the surface of the volume $d\tau$ lead to a change of the particle density as well. Two distinct processes are possible.
4. Spatial gradients influence the net production term because

the diffusive fluxes are unbalanced.

5. External forces change the velocity of particles. Accelerated or decelerated particles leave or enter the volume element in the velocity sub-space (diffusion within the velocity space). Combining all possible gain (G) and loss (L) processes the time derivative of the distribution function can be written as:

$$\frac{\partial f_\alpha}{\partial t} = G_{el} - L_{el} + G_{inel} - L_{inel} + G_{react} - L_{react}$$

$$+ G_{diff} - L_{diff} + G_{ext.F} - L_{ext.F.} \quad \ldots\ldots\ldots\ldots\ldots 1$$

Expressing all gain and loss terms by the appropriate cross-sections and the distribution functions of all kinds of particles, equation 1 changes into Boltzmann's equation.

A simultaneous solution of the total set of Boltzmann's equations gives in principle the distribtuion function f_α of all kinds of particles. The next step is an averaging of microscopic properties over the total velocity interval (as mass, momentum and energy of a single particle) with a final averaging over all kinds of particles. The resulting averaged plasma properties describe macroscopic quantities (such as density, temperature, etc.) which will normally vary as functions of space and time.

In general the method outlined above is not tractable because, without mentioning the mathematical difficulties, the cross sections of most transitions and collisions are not known. Thus one is interested in simplifications where the distribution function is only dependent on a few parameters such as density, temperature, etc. That means the plasma must be not too far from a thermodynamic state. To elucidate this point let us consider the velocity distribution of electrons and heavy particles for arbitrary q and i. Since the elastic collision cross sections between identical particles (and thus the collision rates) in the eV-energy range are very high (compared to the inelastic and reactive one), the gain and loss of particles are dominated by the terms G_{el} and L_{el} in equation 1. Thus the condition

$$G_{el} >> G_{inel},\ G_{react},\ G_{diff},\ G_{ext.F}$$

$$L_{el} >> L_{inel},\ L_{react},\ L_{diff},\ L_{ext.F}$$

is fulfilled in the case of sufficiently small gradients in time and space and small external forces. Then equation 1 reduces to

$$G_{el} \simeq L_{el} \quad \ldots\ldots\ldots\ldots\ldots\ldots \quad 2$$

for elastic collisions between identical particles. It can be proven (by means of the H-theorem) that in this case the velocity distribution is Maxwellian with equal kinetic temperature T for all species s. Moreover, elastic collisions between different kinds of particles occur, the rates of which are included in equation 2. The appropriate cross sections are equally larger than the inelastic one, as can be seen from fig. 1, where for Helium-electron collisions the elastic and inelastic cross-sections (including the ionizing collis-

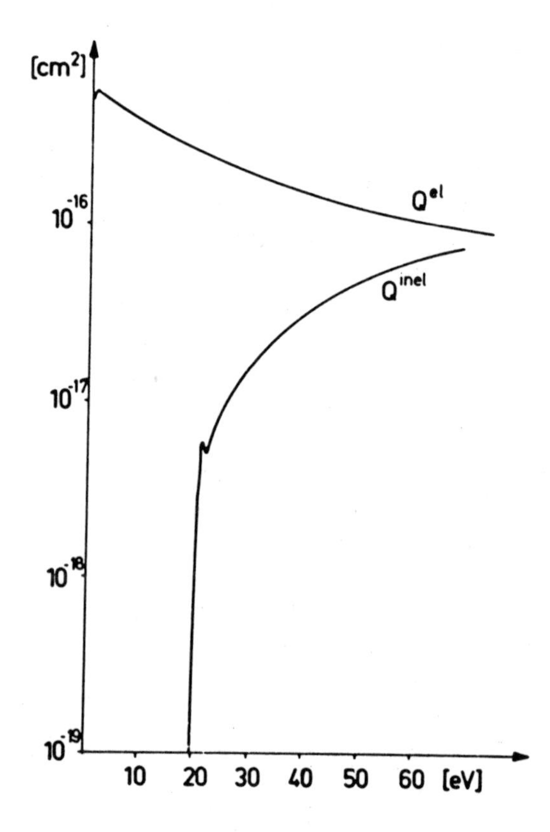

FIG. 1. Elastic and inelastic cross sections for the collision of electrons with He atoms versus electron energy.

ions) are shown as functions of electron energy. However, due to the difference in mass, the effective collision rate between electrons and heavy particles is smaller than between identical particles, or between ions and neutrals provided the number densities are approximately equal. That means the electron gas and heavy particle gas are decoupled. Then loss and gain processes between different kinds of particles are about equal (detailed balance) or they can be neglected in comparison with the rates of collisions between identical particles. In both cases a Maxwellian distribution for all kinds of particles can be assumed with a different kinetic temperature of electrons and heavy particles. This assumption, however, is only justified if the kinetic temperatures of electrons and heavy particles are approximately equal and the electron denisty is so high that the electron-electron encounters provide a Maxwellian distribution.

In arc discharges a Maxwellian velocity distribution with slightly different kinetic temperatures of electrons and heavy particles is established. Then the solution of equation 1 is reduced to the solution of space and time dependent balance equations, which are much easier to handle.

Such simplifications as described above are also used in elastic and reactive processes, diffusion rates and photon loss and production rates. At first we deal with the case of complete equilibrium.

A. Complete Thermal Equilibrium (CTE)
(kinetic, electronic, vibrational, reactive and radiative equilibrium)

The basic assumption of this model is a plasma homogeneous in time, $\partial f_\alpha / \partial t = 0$. External forces derived from a potential may occur inducing a gradient of the distribution function. Then the terms

$$G_{ext.F} - L_{ext.F} + G_{diff} - L_{diff}$$

cancel exactly and all fluxes (e.g. heat and diffusion fluxes) vanish. The velocity distribution function is Maxwellian with equal kinetic temperature for all species. The principle of detailed balancing, previously used for the elastic encounters to derive the Maxwellian distribution, is valid for all conceivable processes and transitions. It then follows from detailed balancing of inelastic processes after velocity averaging ($\bar{G}_{inel} = \bar{L}_{inel}$ e.g. rate of excitative collisions = rate of de-excitative collisions), that the population density of excited levels can be calculated from Boltzmann's formula. The balancing of reactive processes after averaging over velocity leads to Saha's formula (or more generally, to Guldberg and Waages law), which combines the density of particles z or (z+1) times ionized. ($\bar{G}_{react} = \bar{T}_{react}$, e.g. rate of ionizing collisions = rate of three body collision recombination). The bar characterizes the averaging procedure.

Using Boltzmann's equation for the photons including absorption and stimulated and spontaneous emission processes, the complete equilibrium state is found if the rate of absorption equals the rates of spontaneous and induced emissions. The photon distribution function derived from this balance is identical to Planck's radiation law. It ought to be stressed that the temperature occurring in Maxwell's formula (kinetic temperature), Boltzmann's formula (excitational temperature), in Saha's relation (reactional temperature) or in Planck's law (radiative temperature) must be equal. Only with this assumption the detailed balancing $G_{el} = L_{el}, G_{inel} = L_{inel}, G_{react} = L_{react}$ leads to an exact solution of equation 1 with $\partial f_{\alpha}/\partial t = 0$. Then and only then can the plasma be considered to be in complete thermal equilibrium.

In practice it is the equilibrium of back processes that characterize thermal equilibrium rather than the steady state condition, namely $\partial f_{\alpha}/\partial t = 0$

Deviations from thermal equilibrium occur under steady conditions, since gain and loss by the same process do not cancel out. This happens mostly at lower pressures, because there the transition and collision rates are small and three or more body collisions are very unlikely. Arc discharges, for example, are never in complete thermal equilibrium as can be seen by an overall consideration. Electric power is supplied to the plasma via the electrodes, and power is dissipated by radiation and thermal conduction. Because the reverse processes are absent, the principle of detailed balancing is not strictly applicable.

There may, however, be a situation where energy loss by these effects during the mean time between two collisions of particles is only small compared to the total energy exchange between particles. In such a case the equilibrium state is very well approached. Similar considerations applied to detailed processes below lead to a model of incomplete thermal equilibrium, which is very often used in plasma physics.

B. Local Thermal Equilibrium (LTE)
(kinetic, electronic, vibrational and reactive
(collisional or radiative) equilibrium)

In this case the observed distribution functions or mean macroscopic velocities are such that equation 1 is solved only approximately. Small inhomogeneities (besides the gradients due to external potential forces) in space and time may occur. More precisely the change of distribution functions (by external forces) during the mean time between two elastic collisions for identical particles must be small compared to the distribution function itself. A similar estimate must be valid for the excitational (collisional or radiative) and reactive (either collisional or radiative) times of collision. With respect to spatial inhomogeneities the following criterion must be established: the change in the velocity distribution function must be small over the distance of a mean free path for elastic encounters between identical particles. Analogous criteria are valid for the change of population densities over distances of the mean free path between two excitational (either collisional or radiative) or reactive (either collisional or radiative) encounters. Thus the name local equilibrium came into use.

It must be emphasized that not all collision times and mean free paths are subject to the criteria mentioned above, but only those related to dominant processes. In arc plasmas for instance the dominant processes are caused by electron collisions, they lead to excitation as well as to ionization. Thus the mean free path of the photons can be larger than the plasma dimension in contradiction to the validity criteria. In stellar plasmas, however, the radiative processes determine the population of excited and ionized states. Shock wave heating favours the neutrals or ions, in which case heavy particle collisions are mainly of importance.

If all changes of density over mean free path or collision time are small, the remaining loss and gain terms in equation 1 belonging to important processes balance each other, processes with small rates do not. Thus a detailed balancing is only approximately given for processes with high transition rates. The conditions

$$G_{inel} \simeq L_{inel} \text{ and } G_{react} \simeq L_{react}$$

lead to the result that the population of excited and reactive states is described by Boltzmann and Saha equations. The appropriate temperature in the case of arc discharges is the electron temperature. Stellar plasmas with prevailing influence of radiation processes are equally described by Boltzmann or Saha equation but with the radiation temperature as a parameter.

In arc discharges the radiation intensity, however, does not reach Planck's intensity under normal conditions. At least in the visible part of the spectrum most lines and the continuum are emitted from optically thin layers (which means induced emission and absorption rates are small). This assumption may fail with increasing pressure and for very high intensity lines.

Transport processes are allowed in this equilibrium model, the prior condition, however, must be that the transport rates are negligible

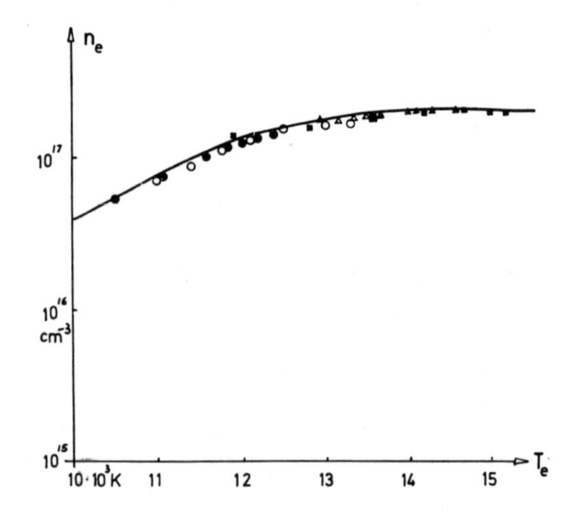

FIG. 2. Electron density versus electron energy in a Krypton discharge (p=1 atm., R=0.15 cm) for different arc currents (■:125A, ▲:100A, Δ:80A, o:60A, ●50A.)

in comparison to the excitation and reaction rates. Diffusion fluxes are sometimes surprisingly high, as can be seen from measurements of the thermal conductivity in the range of dominant dissociation or even ionization reactions.

In high current arcs normally the heavy particles have no influence on excitation and reaction collisions. As mentioned before the kinetic temperature of electrons and ions can differ slightly from each other and it follows from above considerations that it has no influence on the assumption of local thermal equilibrium. Thus the ratio of electron temperature T_e, to heavy particle temperature T_θ may differ from unity,

$$\beta \equiv (T_e/T_\theta) \neq 1 \quad \ldots\ldots\ldots\ldots \quad 3$$

which changes the population density via the equation of state

$$p = n_e kT_e + (n_+ + n_a)kT_\theta \quad \ldots\ldots \quad 4$$

Some transport properties are correlated with the heavy particle temperature, too. However one should not describe the plasma state with $\beta \neq 1$ a priori as a non equilibrium state.

Local equilibrium is often established in arc discharges, as is shown by the following figures. In Fig. 2 electron densities are plotted versus electron temperature in a Krypton discharge, p = 1 atm., R = 0.15 cm, for different arc currents. The solid line was derived from the Saha equation, the experimental points were evaluated assuming a non-equilibrium condition as described in Section 3. The agreement between both results is reasonable. The Krypton plasma is mainly dominated by strong electron collision rates leading to an equilibrium population of all excited levels and thus of the free levels too.

The check for the existence of local equilibrium is often carried out in the following manner (1). The measured absolute intensities of two lines at different stages of ionization and of known transition probabilities are plotted against each other. For optically thin lines the resulting curve is equivalent to plotting the upper levels of the appropriate lines against each other. This curve is easy to calculate assuming local equilibrium and a comparison of measured and calculated data shows the deviation from equilibrium.

Fig. 3 shows as an example the absolute intensity (per cm arc length) of the XeI line 4501Å versus the XeII line 5292Å derived from end-on measurements. The circles characterize the experimental data, the solid line gives the theoretical curve according to the Boltzmann and Saha equations. A systematic difference between the measured and calculated values occurs at high arc currents apparently indicating deviation from local equilibrium. This conclusion would be premature, however, because the XeII line is absorbed inside the plasma and the intensity must be corrected in order to reduce it to the optically thin case. To this end the equation

$$\frac{I(\lambda)}{B_{\lambda}(T_e)} = 1 - \exp - \frac{I(\lambda)_{\theta\tau}}{B\lambda(T_e)} \qquad \ldots\ldots\ldots\ldots \quad 5$$

has to be solved by an iteration procedure (2) for a sufficient number of wavelengths with respect to $I(\lambda)_{\theta\tau}$. Here $\beta_{\lambda}(T_e)$ is the Planck's intensity, $I(\lambda)$ the measured spectral intensity distribution and $I(\lambda)_{\theta\tau}$ is the optically thin measured intensity distribu-

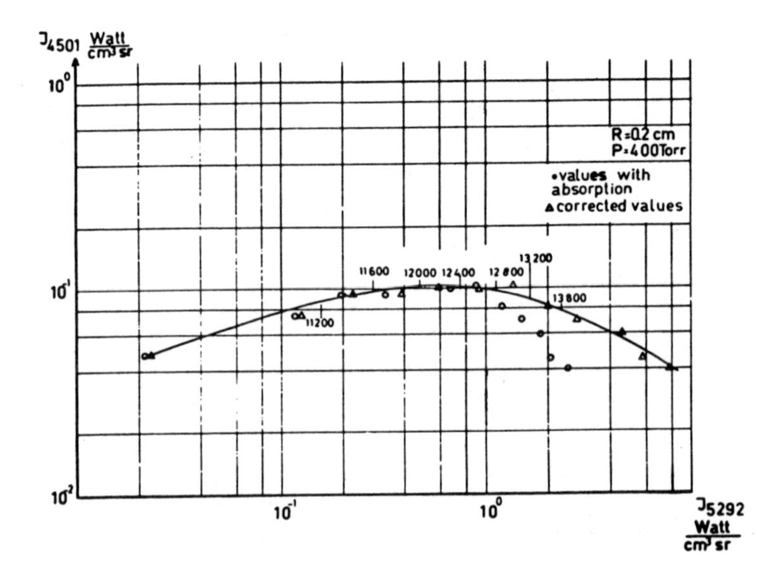

FIG. 3. Intensity of Xe I (4501Å) versus Xe II (5292Å) p = 400 Torr, R=0.2 cm ooooo with absorption ΔΔΔΔ corrected for absorption —— LTE data

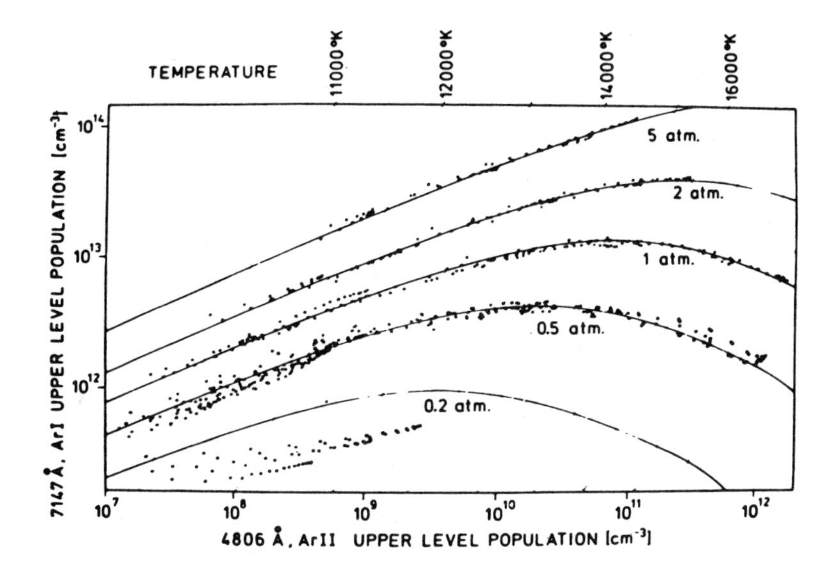

FIG. 4. Population density of upper state of Ar I (7147Å) versus population density of upper state of Ar II (4806Å). For comparison LTE data are plotted. Parameter: arc pressure.

tion. By integration over the wavelength interval the corrected optically thin intensities have been found which are now in good agreement with the local equilibrium data (triangles in Fig. 3). In the xenon discharge for R = 0.2 cm at a pressure of 400 Torr, collision and radiation processes compete with each other at least at 5292Å. By this process the equilibrium situation is improved. Similar measurements were made for arc discharges at different pressures (3). In Fig. 4 the upper level population of ArI 7147Å versus ArII 4806Å is drawn. The solid lines are calculated using local equilibrium assumptions, the data are experimental which are in a very good agreement with the equilibrium results at least for $p > 0.5$ atm. For pressures below 0.2 atm., deviations occur which cannot be explained by absorption processes. Here other mechanisms occur, which are described in the following section.

C. Partial Thermal Equilibrium (PTE)
(Kinetic, electronic and Vibrational)

In arc plasmas the assumption of a Maxwellian velocity distribution with an appropriate kinetic temperature is justified, as was proven

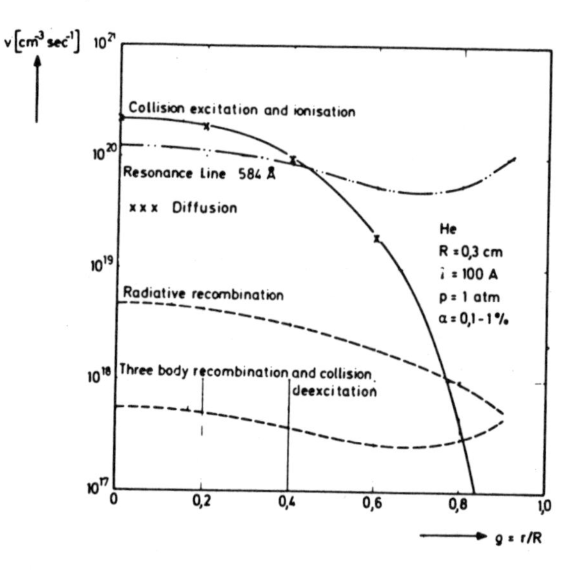

FIG. 5. Collisional and radiative rates for populating and depopulating the ground state of He I 1^1S versus reduced arc radius R=0.3 cm, I=100A, p=1 atm.

by checking the detailed balancing of elastic collisions. Inelastic and reactive rates are much smaller. In many cases collisional and radiative processes and even diffusion phenomena compete with each other. In other words gain and loss of equation 1 do not cancel in a detailed manner. To elucidate this behaviour of an arc plasma Fig. 5 shows a plot of realistic average velocity rates of inelastic processes as a function of the normalized arc radius. The arc is in Helium at atmospheric pressure with a current of 100 A, and radius 0.3 cm. The rates of collisional excitation and ionization are about 10^{20} [cm^{-3} sec^{-1}] on the arc axis. The inverse deexcitation collision processes have rates which are two to three orders of magnitude smaller, while the simultaneously occurring inverse processes of radiative recombination are one to two orders of magnitude smaller. It must be mentioned that the effective transition rates of the resonance line at 584Å are smaller by two orders of magnitude because of strong self-absorption of this line. Summarizing, one can say that within the region of the arc axis essentially all excited and ionized helium atoms are produced by collisions only, with the corresponding inverse processes missing. The situation is completely different close to the wall where the rates of excitation processes are very small and only recombining radiative and collisional processes occur. Here particularly neutrals in the ground state arise.

The magnitude of the rates discussed above leads necessarily to the conclusion that the ground state neutrals vanishing at the axis of the arc must be replaced by new atoms, which come from the wall by diffusion. This is indicated in Fig. 6 by the diffusion flux j_{or} In order to get steady conditions inside the discharge, electrons and ionized atoms have to travel to the discharge walls by ambipolar diffusion. The driving force for both kinds of diffusion fluxes is the pressure gradient of neutrals, electrons and ions respectively, as indicated in Fig. 6.

This example shows clearly that the population density of atoms in the ground state is mainly determined by an equilibrium between collisional excitation and diffusion processes but certainly not by detailed balancing. No equilibrium is established locally in small areas of the arc. Because of the diffusion flux the total arc volume is required to maintain an overall balance of all processes. The ground state population cannot be calculated from the Boltzmann

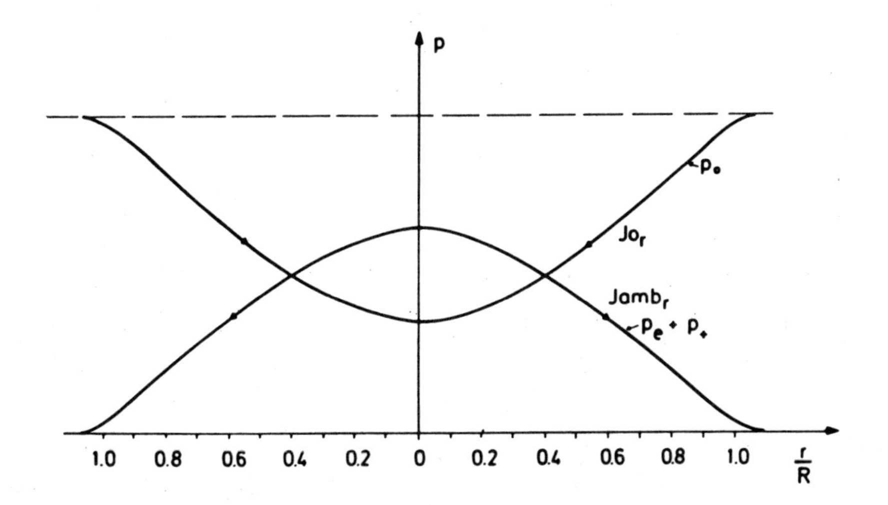

FIG. 6. Radial distribution of partial pressures and diffusion fluxes in a He discharge.

equation and more difficult methods have to be used as described in the next section.

The deviations from local equilibrium as described up to now for the ground state in helium discharge are pronounced for two reasons. First, the cross section for inelastic electron-atom collisions is relatively small, with the corresponding rates small as well. Second, the coefficient of diffusion is large because, since the cross section for charge transfer is small the mean free path for ionizing a ground state helium atom is of the order of magnitude of the arc diameter. Going to higher lying levels the situation is drastically changed. The diffusion coefficient goes down by more than one order of magnitude, the inelastic collision cross sections increase and radiative processes decrease. Thus, as shown in Fig. 7, the excitation rate of electron collisions from the 2^3S state into all states of 3^3P is only twice the rate of the inverse process and the radiative recombination into the 2^3S state is at about the same rate as the collision deexcitation process. Other radiative effects

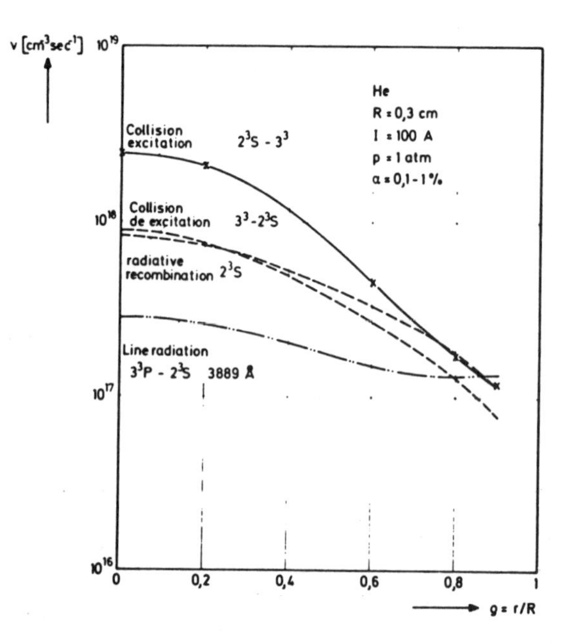

FIG. 7. Collisional and radiative rates for populating and depopulating the He I 2^3S state versus reduced arc radius. R=0.3 cm, I=100A p=1 atm.

are smaller as shown for the 3^3P-2^3S transition. The population density of these excited levels is governed by a collisional-radiative equilibrium. The next higher levels are predominantly populated by electron-atom collision processes. Thus a detailed balancing of collision processes takes place and the Boltzmann formula is applicable. Hence the density of helium atoms in the q-th excited level (with $q > 3$) can be written

$$n_{qPTE} = 1.036 \times 10^{-16} \; (n_e^2/T_e^{3/2}) g_q \; \exp -[(E_j - E_q)/kT_e] \; \ldots\ldots\ldots\ldots \; 6$$

where E_j is the (reduced) ionization energy of Helium, E_q is the excitation energy of the q-th level, k is the Boltzmann constant, g_q is the statistical weight of the q-th level, n_e is the electron density in cm^{-3}, T_e is the electron temperature in ^oK and n_q is in cm^{-3}. Equation 6 is formally equivalent to the Saha equation for the density of an excited level, but since the formula is not applicable down to the lowest levels it cannot be used to calculate the electron density. In order to get at least some connection with the equilibrium situation in a quite formal manner the ratio

$$b_q \equiv \frac{n_q}{n_{qPTE}} \quad \ldots\ldots\ldots\ldots\ldots \; 7$$

is introduced. For levels with $q > 3$ one obtains $b_q = 1$. Thus according to equation 6 and 7 the total density of neutrals is found to be

$$n_a = \sum_{q=0}^{q_{max}} b_q \; n_{qPTE}$$

$$= n_{aLTE}(1/Z_a) \sum_{q=0}^{q_{max}} b_q g_q \; \exp -(E_q/kT_e) \; \ldots\ldots \; 8$$

where Z_a is the partition function for the neutrals. Knowing b_q and using the condition of quasineutrality the electron density is obtained from the equation of state (equation 4).

Since $b_q = 1$ for some higher excited levels this equilibrium is called partial equilibrium. Equation 6 is often very useful for temperature measurements from absolute line intensities, when n_e is known. Equation 8, however, is only of interest if the coefficients b_q are known. The b_q-factors can be measured or calculated. Fig. 8 shows a plot of b_q against the reduced arc radius r/R for an He discharge for p = 1 atm., I = 140A, and R = 0.3 cm. These data were calculated in accordance with the theory in the next section. The b_q factor of the ground state 1 S is about 230 (for this example) on the arc axis and decreases towards the wall. In other words, the ground state density near the arc axis is much higher than expected from a pure thermodynamic consideration. The reason for this phenomenon here is the dominating influence of collisional excitative and diffusive processes as discussed earlier, leading to a diminished rate of ionization and to a higher number of neutral atoms. At the wall the electron density is high due to ambipolar diffusion effects and thus the b_q-factors are smaller. Levels with a main

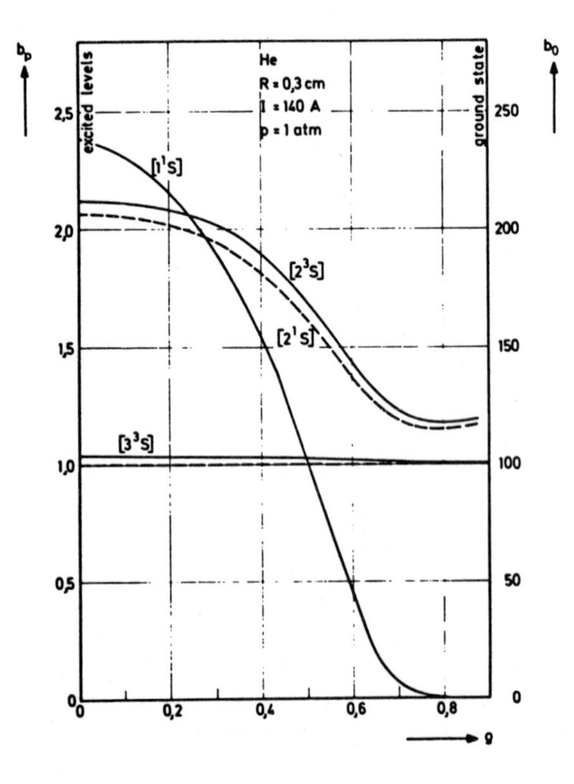

FIG. 8. Non-equilibrium factor b_q for different He I levels versus reduced arc radius. R=0.3 cm, I=140A, p=1 atm.

quantum number larger than 3 are in partial equilibrium all over the arc cross section, this limit however moving towards higher levels if the pressure inside the discharge is decreased.

III. THERMAL-NONEQUILIBRIUM FOR EXCITATIVE AND REACTIVE PROCESSES

A. The Basic Equations

In this section we discuss calculation of b_q for levels which are not in partial equilibrium. From statements up to now it is clear that b_q is not determined by local variables, and that the total temperature and density distributions must be known if diffusion fluxes exist.

The usual way to derive the appropriate equations is to multiply equation 1 with velocity moments Q_α (e.g. 1, V_α, V_α^2) belonging to particles of kind α. After integration over velocity space the transport equations are derived, which lead to a relation between macroscopic quantities. Particle balance, momentum balance and energy balance conditions are obtained for $Q_\alpha=1$, V_α and V_α^2 respectively, which describe the space-time behaviour of an arc discharge.

If the velocity distribution function of the different species is developed around a Maxwell distribution, equation 1 can also be used to derive the transport coefficients.

In the following, we confine ourselves to the case of a steady arc plasma with radial dependence of the macroscopic quantities. Then the equation of continuity for the k-th excited level can be written as

$$\frac{1}{r}\frac{d}{dr} r j_{kr} = \sum_{q=0}^{k-1} n_q n_e C_{qk} + \sum_{q=k+1}^{q_{max}} n_q (n_e F_{qk} + A_{qk}\Lambda_{qk})$$

$$- \sum_{q=0}^{k-1} n_k (n_e F_{kq} + A_{kq}\Lambda_{kq}) - \sum_{q=k+1}^{q_{max}} n_k n_e C_{kq}$$

$$+ n_+ n_e (R_k \Lambda_k + n_e Q_k) - n_e n_k S_k \qquad \text{............... 9}$$

where

C_{qk}, F_{qk} are rate coefficients for collisional excitation and deexcitation,

$R_k, n_e Q_k$ are rate coefficients for radiative and collisional recombination,

S_k is the rate coefficient for collisional excitation

A_{qk} is the optical transition probability,

Λ_k, Λ_{qk} are reduction coefficients, which take into account absorption of the continuum and of line radiation,

and n_q is the density of the atoms in level q.

The particle diffusion flux in level k is only of importance for ground state atoms, at least in the pressure range discussed here. Kinetic theory gives the expression

$$j_{or} = -(3/8)[(n_o + n_+)\Omega_{a+}^{(1,1)} m_G]^{-1} \frac{dP_a}{dr} \quad \ldots\ldots\ldots\ldots\ 10$$

assuming

$$j_{kr} \simeq 0 \quad \ldots\ldots\ldots\ldots\ldots\ldots\ 11$$

Here m_G is the mass of He atoms, $\Omega_{a+}^{(1,1)}$ is a particular Ω- integral for atom-ion charge transfer and p_a is the partial pressure of neutral atoms.

Equation 9 is easily reduced to the conditions of local equilibrium by balancing the rates of direct and inverse processes, e.g.

$$n_e n_q C_{qk} \simeq n_e n_k F_{kq} \quad \ldots\ldots\ldots\ldots\ 12$$

This equation coincides with the Boltzmann equation.

If the diffusion term in equation 9 is totally neglected the remaining system of equations is a kind of generalized Corona equation Measurements show that this type of equilibrium relation has just as restricted a use for constricted arc plasmas, as Saha's equation discussed above.

The solution of equation 9 with the simplification in equation 11 and equation 4 the equation of state, is only possible if the temperature profile of electrons and heavy particles is known, because the rate coefficients are functions of electron temperature and, Λ and $\Omega_{a+}^{(1,1)}$ depend on the gas temperature. Two different methods were chosen to evaluate the rate equation.

a. In the first method, starting with measured absolute line intensities of optically thin lines with known transition probabilities the density profile of upper lines, $n_q(r)$, can be derived. The knowledge of $n_q(r)$ is sufficient to solve equation 9 provided all rate coefficients are known. This method in principle agrees with the temperature evaluation in LTE plasmas, where the temperature dependence of line intensities can be easily found. The heavy particle temperature is derived from the energy balance of the electron gas

$$\sigma E^2 = \frac{1}{r}\frac{d}{dr} r q_{er} + (4m_e/m_G)k(T_e - T_G)\ \xi_{er} + U_{inel} \quad \cdots\cdots\cdots\cdots\cdots \quad 13$$

where σ is the electrical conductivity, E is electrical field, q_{er} is heat flux of the electron gas, ξ_{er} is the elastic collision frequency, and U_{inel} is the energy loss (or gain) by inelastic collisions. Equation 13 must be solved simultaneously with equation 2. Results derived from this procedure are called "experimental" in the following.

b. In the second method a complete theoretical description of the arc plasma without use of any measured data requires some additional information about the thermal structure of the plasma. One can get this information from the energy balance of the plasma which is

$$\sigma E^2 = \frac{1}{r}\frac{d}{dr} r q_r + \operatorname{div} \vec{j}_{ph} \quad \cdots\cdots\cdots\cdots \quad 14$$

where q_r and $\vec{j}_{ph}$ are thermal heat flux transport by particles and photons respectively. The results of a simultaneous solution of equations 9, 13 and 14 including 4 are called "energy balance" in some figures below.

B. Experimental Verification of the Non-Equilibrium Model

There are a number of direct and indirect experimental methods to verify the reliability of the model derived. Such experiments are of importance because there is an uncertainty in the rate coefficients and cross sections. One of the simplest methods which come to mind is the measurement of the arc characteristic, which depends sensitively on the equilibrium conditions in the arc plasma.

Measurements of the E-I characteristic can be compared with the theoretical data. This has been done elsewhere (4,5) and the comparison gave good agreement.

Another direct diagnostic tool is light scattering (6). In He discharges at atmospheric pressure the scattering factor α is about 1, and so the electron temperature as well as the electron density can be derived with relatively high accuracy. Fig. 9 shows a comparison of different temperature measurements for a He discharge.

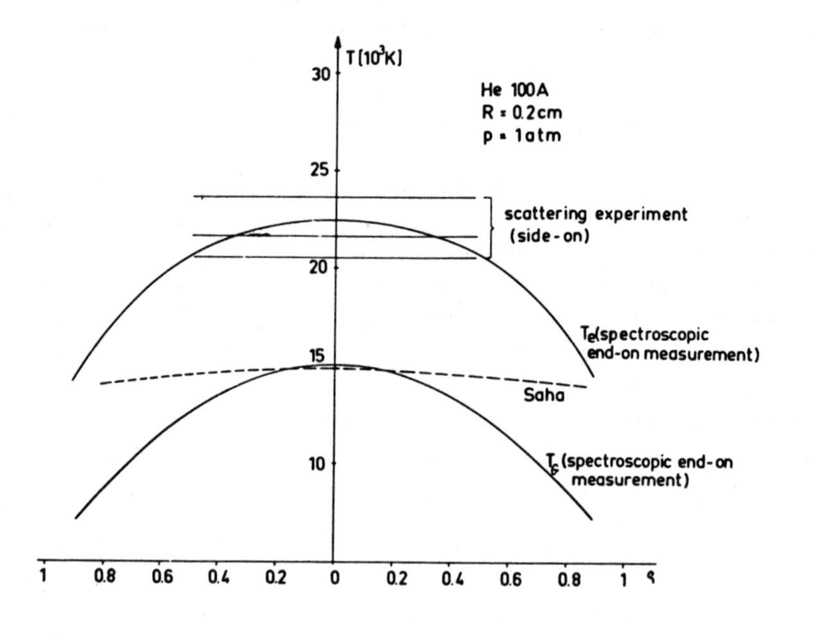

FIG. 9. Comparison of different temperature measurements in an He discharge (p=1 atm, R=0.2 cm, I=100A).
Straight bars: Scattering experiment
Solid Line: Spectroscopic measurement of electron and gas temperature using method A (see text)
Dashed Line: Saha evaluation of measured absolute intensities.

The horizontal bars indicate the results of three independent scattering experiments as well as the space resolution. Results of radial temperature profiles derived by method A using end-on absolute intensity measurements are also shown. The upper, solid curve gives the electron temperature, and the lower one gives the gas temperature. The dashed curve is calculated in the standard way using an LTE evaluation. The agreement of the fictitious Saha temperature with the real gas temperature in the arc axis is only accidental. The discrepancy between temperatures derived from a

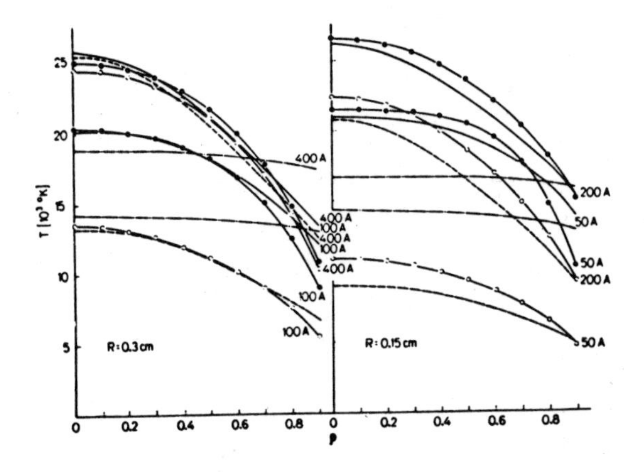

FIG. 10. Comparison of calculated and measured temperature profiles.
_._._._. LTE evaluation of absolute line intensities
_._._._. electron temperature according to method B
-o-o-o-o gas temperature according to method B
———— electron temperature according to method A
- - - - gas temperature according to method A

In the case of R=0.3 cm(left) the line intensity of He I 4713Å was used, and for R=0.15 cm(right), the line intensity of He I 5015Å.

pure LTE evaluation and according to method A is evident, as can be seen from Fig. 10. Here spectroscopic measurements of absolute line intensities of HeI 4713Å and HeI 5015Å were used to derive the temperature profiles of electrons (solid line) and heavy particles (dashed line) via the rate equations. According to method B (using the energy balance), the same curves are derived, the dashed curves with full circles for the electron temperature and the dashed curves with open circles for the gas temperature. The LTE curve (dashed-dotted) is far from reality. The figure on the left gives data for a He arc with radius 0.3 cm, I = 100A and 400A, and the figure on the right is for an arc with radius 0.15 cm, I = 50A and

200A.

It is interesting to consider the electron density profile, because the diffusion process may alter the density more directly than the temperature. Fig. 11 gives a comparison of calculated data using method B (solid line)with direct density measurements from line widths (triangles for HeI 5015Å, circles for HeI 4713Å). Scattering

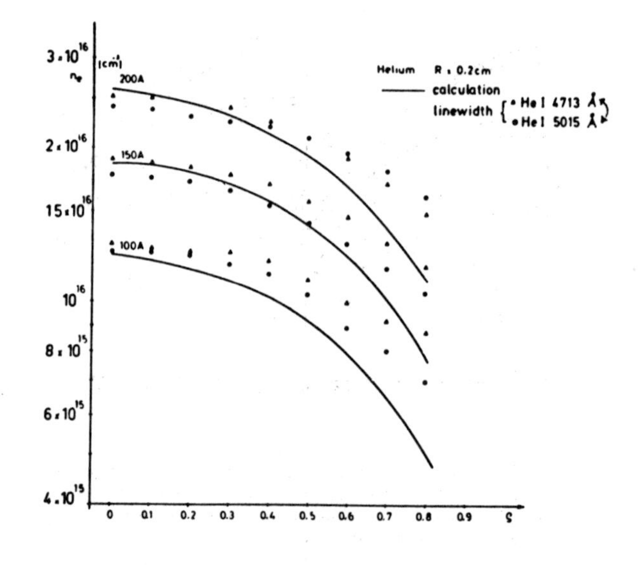

FIG. 11. Radial distribution of electron density
solid line: calculated profiles (method B)
circles: from line width of He I 4713 Å
triangles: from line width of He I 5015 Å

results on the arc axis (7) agree with these data within 10%. The deviations close to the wall are not fully understood. In order to visualize the non-equilibrium case, electron densities are plotted against electron temperature in Fig. 12 as already done in Fig. 2 for a Krypton discharge. For the helium discharge (R = 0.3 cm, p = 1 atm.) no unique curve for n_e (or T_e) results. The Saha curve (dashed-dotted) as well as a generalized Corona equation (dashed) do not describe the realistic density profile. The solution of the set of basic equations following method A (solid line) or B (open circles) as well as line width measurements from a small admixture

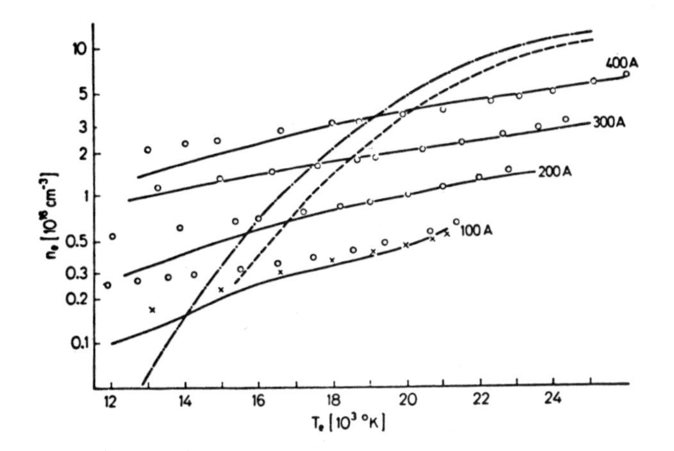

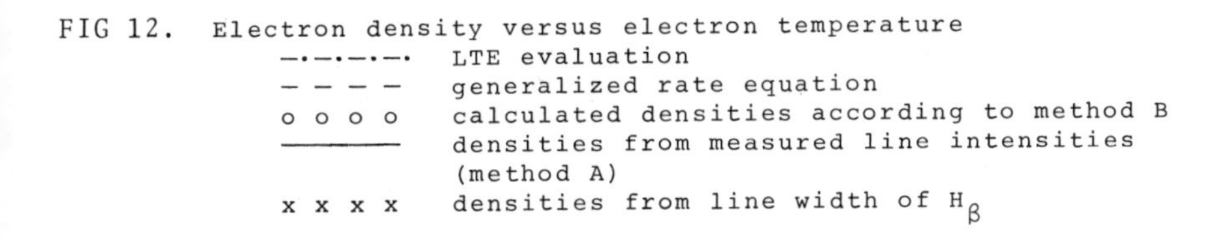

FIG 12. Electron density versus electron temperature
—·—·—·— LTE evaluation
— — — — generalized rate equation
o o o o calculated densities according to method B
———— densities from measured line intensities (method A)
x x x x densities from line width of H_β

of H_2 are in good agreement. The realistic density curves cross the Corona curve and the Saha curve at one point each. A volume element situated at this point suffers a considerable diffusion flux, but the fluxes through the surfaces at r and r + dr are approximately equal, thus collision and radiation processes balance exactly. Interferometric studies of He discharges support the non-equilibrium model in a similar way (8).

Similar non-equilibrium effects were also studied by others (9,10) and in other gases (11). In Fig. 13 electron density profiles of a D_2 discharge at 450 Torr pressure are plotted for I=55A and R=0.2 cm. Densities from an LTE evaluation measuring absolute intensities of D_α and D_β do not agree with scattering results (7) and data derived from measurements of the continuum, thus indicating non-equilibrium effects. As in He the real temperature of the D_2 discharge is higher than expected from an LTE evaluation.

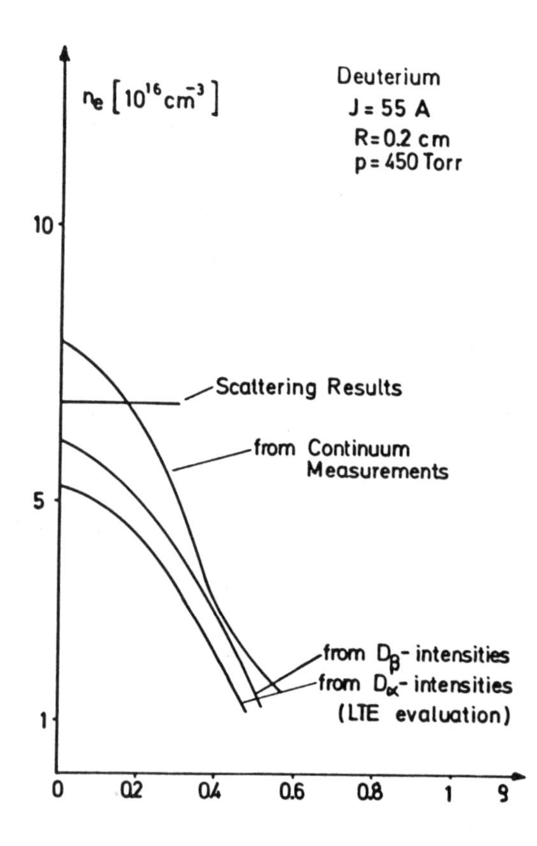

FIG. 13. Comparison of electron density profiles in a D_2 discharge (I=55A, r=0.2 cm, p=450 Torr) using different experimental methods.

IV. NON-EQUILIBRIUM DUE TO STRONG HEAT FLUXES

In section III it was shown that strong diffusion fluxes inside an arc disturb the LTE condition so that the population density no longer has a Boltzmann distribution. The velocity distribution, however, remains Maxwellian because of the dominating influence of elastic collisions. With decreasing pressure the mean free path for elastic collisions increases and even deviations from the Maxwellian distribution occur, for instance due to external electric fields or inelastic collisions.

A large disturbance of the velocity distribution is also introduced in the vicinity of the walls of the discharge vessel. Particles moving towards the wall have a "higher temperature" than particles leaving the wall. The disturbance is large in a region a few mean paths away from the wall. In order to estimate the effect one has to compare the directed particle flux $\vec{j}_D$ with the isotropic flux $n\bar{V}_{th}$ or in a uniform gas without diffusion one must compare the directed heat flux $\vec{q}$ with the iostropic flux $nkT\,v_{th} \simeq p\bar{v}_{th}$. Here n is the number density and $\bar{v}_{th}$ is the mean thermal speed. If the ratio

$$\gamma = (q/p\bar{v}_{th}) \quad \ldots\ldots\ldots 15$$

is not negligible compared to unity, t' Maxwellian distribution is strongly affected. The ratio γ is not small for heavy gases in high power arcs where a $q \simeq 20 kW/cm^2$ in front of the wall even at atmospheric pressures. The situation is much worse in fusion reactors, because there the pressure is much lower than in atmospheric arcs.

The starting point for the theoretical investigation is again Boltzmann's equation. According to Lees and Liu (12) an "ansatz" of the form

$$f(\vec{v}) = \begin{cases} n_+ \quad (m/2\pi kT_+)^{3/2} \exp -(mv^2/2kT_+) & o \le v_z < \infty \\ n_- \quad (m/2\pi kT_-)^{3/2} \exp -(mv^2/2kT_-) & -\infty \le v_z \le o \end{cases} \quad \ldots. 16$$

is quite useful. Here n_+, n_-, and T_+, T_- are the densities and temperatures of particles travelling away from the wall (+z direction) or towards the wall (-z direction) with $T_- > T_+$. This so-called bimodal distribution (see Fig. 14) is of importance in heat transfer problems of ordinary gas-dynamics. It makes possible calculation of transport properties of dilute gases in the range between the Knudsen regime and the Chapman-Enskog regime.

It is useful to apply this theory to the case of a gas mixture (plasma) in order to find the temperature and density distribution close to the discharge wall. The one dimensional plane case for a uniform gas will be treated first. The macroscopic averages expressed by n_+, n_- and T_+, T_- are

$$n = \iiint_{-\infty}^{\infty} f\, d^3\vec{v} = 0.5(n_+ + n_-)$$

$$\bar{v}_z = \iiint_{-\infty}^{\infty} f\, v_z\, d^3\vec{v} = (2k/\pi m)^{\frac{1}{2}} (n_+T_+^{\frac{1}{2}} - n_-T_-^{\frac{1}{2}}) \quad \ldots .\ 17$$

$$(\tfrac{3}{2})kT = (m/2n) \iiint_{-\infty}^{\infty} f(\vec{v}-v_z e_z) d^3\vec{v} = (3k/2)(n_+T_+ + n_-T_-)/(n_+ + n_-)$$

$$p = nk\bar{T}$$

$$q_z = (2k/\pi m)^{\frac{1}{2}}(n_+T_+^{3/2} - n_-T_-^{3/2})$$

They are connected by the transport equations,

$$\frac{\partial}{\partial z} n\bar{v}_z = 0 : n_+T_+^{\frac{1}{2}} = n_-T_-^{\frac{1}{2}} \quad \text{(equation of continuity)} \quad \ldots\ldots\ 18$$

$$\frac{\partial}{\partial z} n\, \overline{\vec{v}v_z} = 0 : nT = \text{Constant} \quad \text{(momentum equation)}$$

$$\frac{\partial}{\partial z} n\, \overline{(\vec{v}^2/2)v_z} = 0 : q_z\ \text{Constant} \quad \text{(energy balance)}$$

$$\frac{\partial}{\partial z}(n_+T^2 + n_-T_-^2) = I$$

$$= (2m^2/5k^2) \iint_{-\infty}^{\infty} (v'^2v_z' - v^2v_z) f_1 f |\vec{v}_1 - v| \sigma d\Omega d^3\vec{v}_1 d^3\vec{v}$$

The last formula is a transport equation belonging to the higher moment $(m/2)\bar{v}^2\vec{v}$. The collision integral I is very difficult to handle for arbitrary interparticle potentials, evaluations for hard spheres and Maxwellian molecules only being available (12,13). Thus the distribution functions f and f_1 were developed into Hermitian polynomials in analogous to Grad's theory in order to get simple expressions for the integral

$$f = n(m/2\pi kT)^{3/2}\ [\exp - (mv^2/2kT)]\ [1 - mv_zq_z/nk^2\bar{T}^2)(1 - m\vec{v}^2/\ (5k\bar{T})] \quad ..19$$

This distribution function is plotted in Fig. 14 as a solid line. The bimodal distribution fits very well especially at the wings of

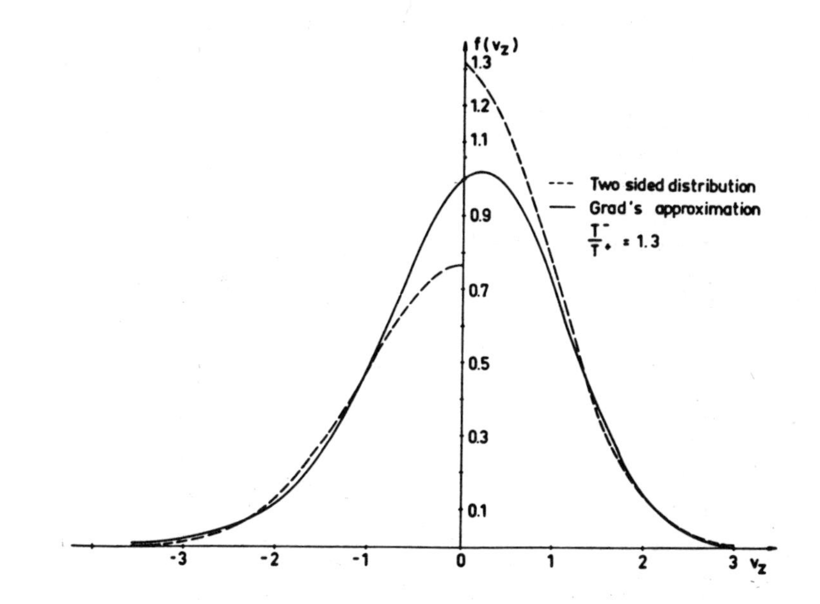

FIG. 14. Approximation of the bimodal distribution by Grad's polynominals.

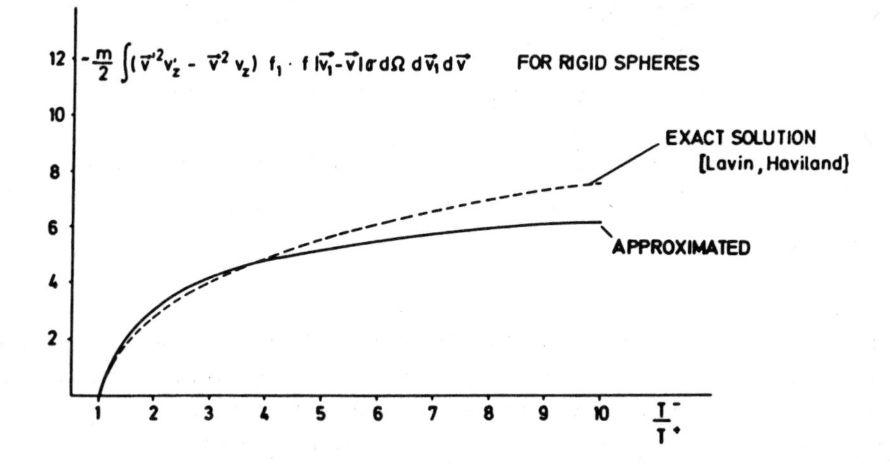

FIG. 15. Evaluation of the collision term for rigid spheres (---) using the bimodal distribution and the approximation with Hermite's polynomials.

the curve. Using equation 18 the integral I can be expressed by the well known Ω integrals. Fig. 15 shows a comparison of the value of I as a function of T_-/T_+ according to Lavin and Haviland (13) using the bimodal distribution (dashed line) and the approximated curve from equation 16. The agreement is fair up to large ratios T_-/T_+.

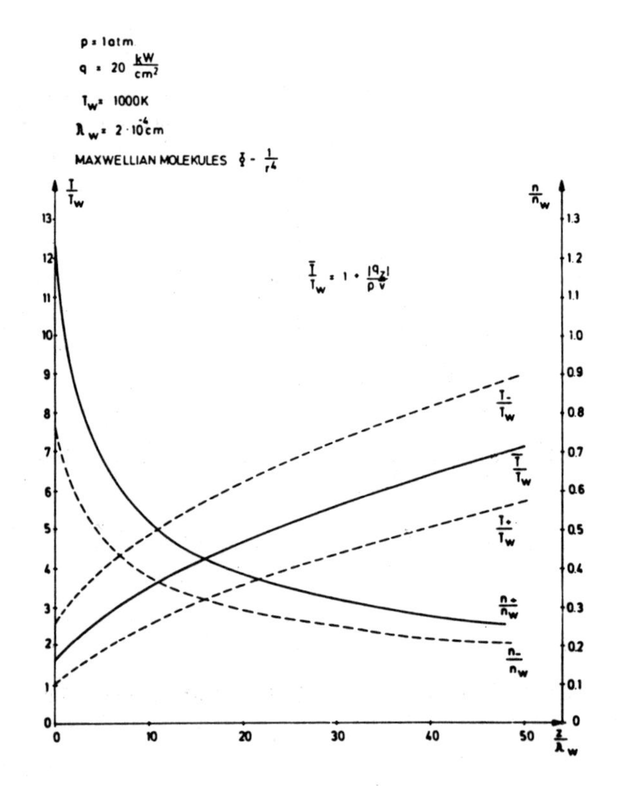

FIG. 16. Temperature and density distribution close to the wall of a Hydrogen discharge at atmospheric pressure. The wall distance is normalized to the mean free path.

The set of equations 17 and 18 can be solved simultaneously assuming a constant heat flux q_z flowing to the wall with vanishing mass flow. After normalizing all temperatures to the wall temperature T_W, one obtains the average gas temperature as a function of z as

$$T(z) = \frac{\bar{T}(z)}{T_W} = (T_-(z)/T_W)^{\frac{1}{2}}(T_+(z)/T_W)^{\frac{1}{2}} \quad \ldots\ldots\ldots\ldots\ldots\ldots \quad 20$$

and the normalized mean gas temperature as

$$T_G/T_W = T(0) = 1+q_z/(pv_{th}) = 1 + \gamma \quad \ldots\ldots\ldots\ldots\ldots \quad 21$$

where γ is defined in equation 15.

Fig. 16 shows plots of density and temperature as a function of distance from the wall for a hydrogen discharge at 1 atm. pressure, and distance being normalized to the mean free path. The average gas temperature at the wall is higher than the real wall temperature. With increasing distance from the wall the relative differences of $T_- T_+$ and $\overline{T}$ reduce and the Chapman-Enskog theory is applicable.

This can be seen clearly in Fig. 17, where the average temperature is plotted against distance from the wall for a deuterium discharge with q_z=10 W/cm^2 and n_W = 10^{14}cm^{-3}. The Chapman-Enskog theory is quite realistic over a distance a few mean free paths away from the wall. These results were obtained for a plane geometry.

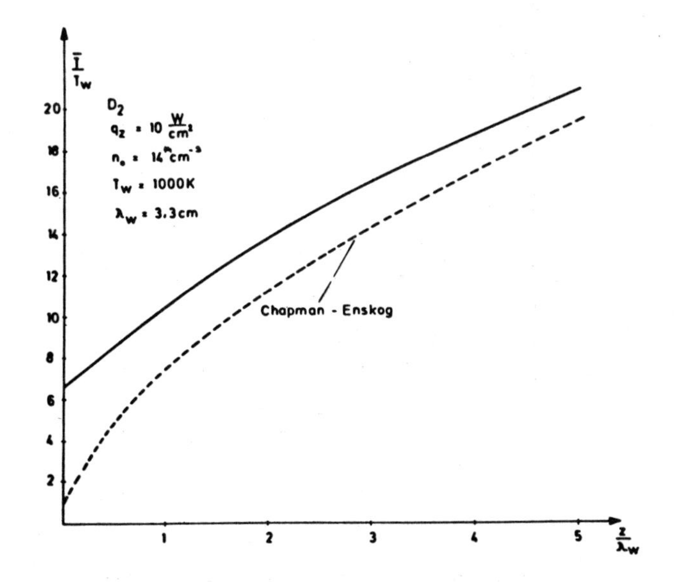

FIG. 17. Temperature distribution close to the wall of a Deuterium discharge. Particle density 10^{14}cm^{-3}. For comparison, results according to the Chapman-Enskog method are given.

Finally, the cylindrical case was investigated. Here the thermal flux depends on the radius and for a weakly ionized discharge with LTE conditions the energy balance becomes

$$(2k^3/\pi m)^{\frac{1}{2}} \frac{1d}{rdr} r(n_+T_+^{3/2} - n_-T_-^{3/2}) = \sigma(\overline{T})E^2 \quad \dots\dots\dots\dots\dots \quad 22$$

where

$$-K \frac{d}{dr} (n_+T_+^2 + n_-T_-^2) = n_+T_+^{3/2} - n_-T_-^{3/2}$$

and $\quad K = (75/64)\ (2\pi k/m)^{\frac{1}{2}}[\Omega^{(2,2)}(n_+ + n_-)]^{-1}$

The solution of equation 22 for a He discharge with p = 0.1 atm. and R = 0.2 cm for four different temperatures on the arc axis and fixed wall temperature is shown in Fig. 18 (solid line). The dashed lines are solutions of the usual energy balance (Elenbaas-Heller equation). The largest deviation between dashed and solid curves

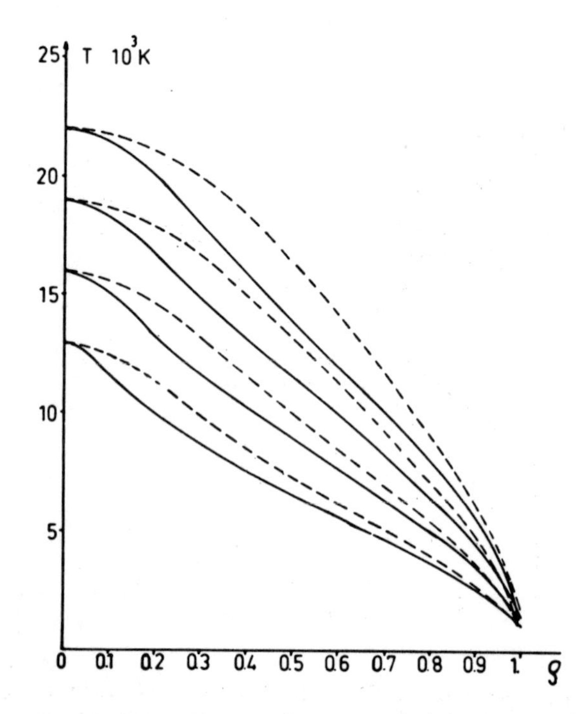

FIG. 18. Comparison of radial temperature distribution in a He discharge (R=0.2 cm, p=0.1 atm) derived from LTE (- - - -) and assuming a bimodal distribution (———).

occur where the heat flux is a maximum. Strong heat fluxes may thus influence the thermal behaviour of the discharge plasma.

V. SUMMARY

Because all non-equilibrium calculations are very tedious, one is interested in simple criteria to check whether LTE is established. Summarizing the ideas of Section 2, the LTE requires that

1. The Velocity distribution of particles is Maxwellian (not necessarily with equal temperatures for the different species).
2. Inelastic collision processes must dominate all radiative contributions.
3. Mean free path of a neutral particle must be small compared with the linear dimension of the arc. In Fig. 19 the ratio of arc

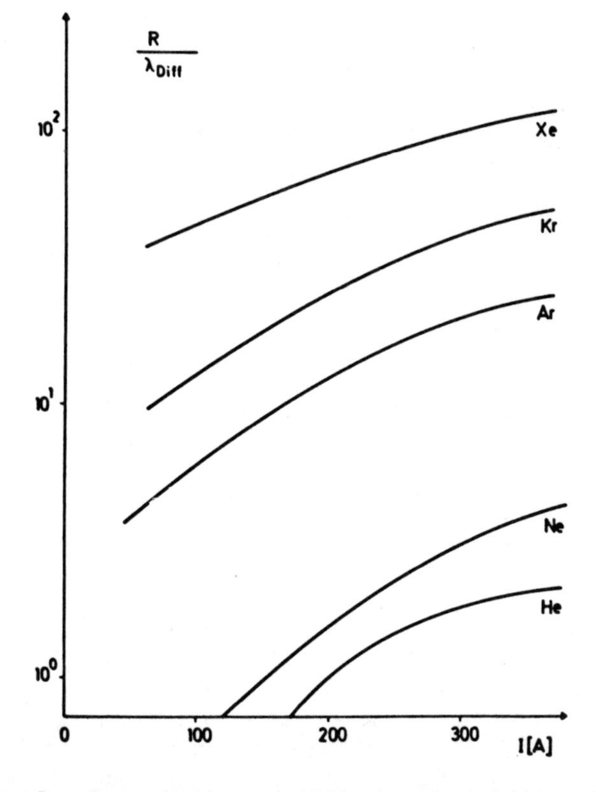

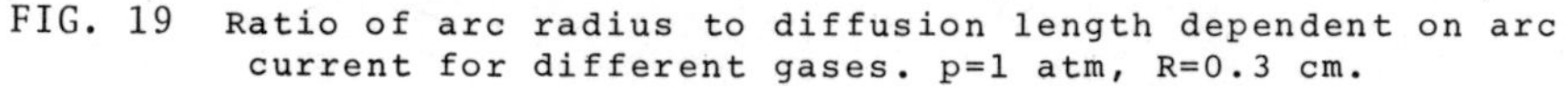

FIG. 19 Ratio of arc radius to diffusion length dependent on arc current for different gases. p=1 atm, R=0.3 cm.

diameter R to diffusion length γ is plotted against arc current (p = 1 atm). R/γ is only small for the rare gases Xe, Kr and Ar in the low current regime. Helium and even Ne are more difficult to handle and show a non-equilibrium behaviour.

Plasma properties are influenced by a strong heat flux which is dominant close to the wall within a distance of a few mean free paths. The average gas temperature $\overline{T}_G$ is higher than the wall temperature T_W. The temperature ratio is fairly accurately given by

$$\overline{T}_G/T_W = 1 + q_z/(p\bar{v}_{th}) = 1 + \gamma$$

where q_z is the heat flux, p is the discharge pressure and $\bar{v}_{th}$ is the average particle velocity. If $\gamma \ll 1$ the temperature change at the wall can be neglected.

VI. ACKNOWLEDGEMENTS

I would like to take this opportunity to sincerely thank Prof. B. Ahlborn of the Physics Department of the University of British Columbia, Canada for numerous discussions and for reading this manuscript.

REFERENCES

1. Richter, J., Phenomena in Ionized Gases, Oxford 1971, Invited papers.
2. Hackmann, J.H. Michael, Uhlenbusch, J., Z. Phys. 250, 207 (1972)
3. Schumaker, J.B. and Popenoe, C.H., Journal of Research of NBS, A, 76A, 2, 71 (1972).
4. Uhlenbusch, J. and Fischer, E., Proc. IEEE, 59, 578 (1971).
5. Uhlenbusch, J., Fischer, E. and Hackmann, J., Z. Phys. 238, 404 (1970); 239, 120 (1970).
6. Gieres, G., Kempkens, H., and Uhlenbusch, J., Atomkernenergie 19, 205 (1972).
7. Gieres, G., Thesis, University Aachen (1973).
8. Baum, D., Hackmann, J. and Uhlenbusch, J., to be published.
9. Krüger, C.H., Phys. Fluids, 13, 1737 (1970).
10. Bergstedt, K. Naturforsch, Z., A24, 299 (1969).
11. Wiese, W., VI Yugosl. Symposium on Physics of Ionized Gases, Miljevac by Split (1972).
12. Lees, L. and Liu, C.Y., Phys. Fluids 5, 1137 (1962).
13. Lavin, M.L. and Haviland, J.U., Phys. Fluids 5, 274 (1962).